Building Construction: Principles, Materials and Systems

Building Construction: Principles, Materials and Systems

Charles Bale

Larsen & Keller
www.larsen-keller.com

Building Construction: Principles, Materials and Systems
Charles Bale
ISBN: 978-1-64172-655-9 (Hardback)

Larsen & Keller
Published by Larsen and Keller Education,
5 Penn Plaza,
19th Floor,
New York, NY 10001, USA

Cataloging-in-Publication Data

Building construction : principles, materials and systems / Charles Bale.
p. cm.
Includes bibliographical references and index.
ISBN 978-1-64172-655-9
1. Building. 2. Construction industry. 3. Building materials. 4. Structural engineering. I. Bale, Charles.
TH145 .B85 2022
690--dc23

For more information regarding Larsen and Keller Education and its products, please visit the publisher's website www.larsen-keller.com

Table of Contents

Preface

The process of constructing an infrastructure or building is known as construction. It deals with the planning, designing and financing of the construction project. The various sectors into which construction can be classified are buildings, infrastructure and industrial. The construction of buildings is divided further into commercial or institutional buildings and residential buildings. Some of the steps which are a part of the process of constructing buildings are developing floor plans, excavating the foundation and building the main load bearing structure. Finally, there is an interrelation and overlap of different fields in order to ensure that the building is structurally sound, financially possible to build and fulfill the legal requirements. This book provides significant information of this discipline to help develop a good understanding of building construction and related fields. Those in search of information to further their knowledge will be greatly assisted by this book.

To facilitate a deeper understanding of the contents of this book a short introduction of every chapter is written below:

Chapter 1- The process of constructing buildings or adding structure to real estate is known as building construction. This is an introductory chapter which will introduce briefly all the significant aspects of building construction such as the modern building practices and construction site safety.

Chapter 2- The different materials which are used for the purpose of construction are known as building materials. Some of the different types of building materials are mud and clay, fabric and natural construction materials. The chapter closely examines these types of building materials as well as their properties to provide an extensive understanding of the subject.

Chapter 3- Brick refers to the building material which is used to make pavements, walls and several other elements in masonry construction. The form of construction where bricks are placed together in a systematic manner and mortar is added to the gaps is known as brick masonry. This chapter discusses brick masonry in detail as well as the different types of bricks.

Chapter 4- The construction which makes use of stones that are bonded together by mortar is known as stone masonry. The structural material consisting of a hard and chemically inert particulate substance which is bonded together by cement and water is called concrete. The diverse applications of stones and concrete in construction have been thoroughly discussed in this chapter.

Chapter 5- Frame construction involves the fitting together of pieces in order to give shape or support to a structure. Primarily wood or steel are used as framing materials. The topics elaborated in this chapter will help in gaining a better perspective about wood and steel frame construction as well as the different properties of wood and steel.

I owe the completion of this book to the never-ending support of my family, who supported me throughout the project.

Charles Bale

1

Building Construction: An Introduction

The process of constructing buildings or adding structure to real estate is known as building construction. This is an introductory chapter which will introduce briefly all the significant aspects of building construction such as the modern building practices and construction site safety.

Construction is also called building construction, the techniques and industry involved in the assembly and erection of structures, primarily those used to provide shelter.

Construction is an ancient human activity. Constructed shelters were one means by which human beings were able to adapt themselves to a wide variety of climates and become a global species.

Human shelters were at first very simple and perhaps lasted only a few days or months. Over time, however, even temporary structures evolved into such highly refined forms as the igloo. Gradually more durable structures began to appear, particularly after the advent of agriculture, when people began to stay in one place for long periods. The first shelters were dwellings, but later other functions, such as food storage and ceremony, were housed in separate buildings. Some structures began to have symbolic as well as functional value, marking the beginning of the distinction between architecture and building.

The history of building is marked by a number of trends. One is the increasing durability of the materials used. Early building materials were perishable, such as leaves, branches, and animal hides. Later, more durable natural materials—such as clay, stone, and timber—and, finally, synthetic materials—such as brick, concrete, metals, and plastics—were used. Another is a quest for buildings of ever greater height and span; this was made possible by the development of stronger materials and by knowledge of how materials behave and how to exploit them to greater advantage. A third major trend involves the degree of control exercised over the interior environment of buildings: increasingly precise regulation of air temperature, light and sound levels, humidity, odours, air speed, and other factors that affect human comfort has been possible. Yet another trend is the change in energy available to the construction process, starting with human muscle power and developing toward the powerful machinery used today.

The present state of construction is complex. There is a wide range of building products and systems which are aimed primarily at groups of building types or markets. The design process for buildings is highly organized and draws upon research establishments that study material properties and performance, code officials who adopt and enforce safety standards, and design professionals who determine user needs and design a building to meet those needs. The construction process is also highly organized; it includes the manufacturers of building products and systems, the craftsmen who assemble them on the building site, the contractors who employ and coordinate the work of the craftsmen, and consultants who specialize in such aspects as construction management, quality control, and insurance.

Construction today is a significant part of industrial culture, a manifestation of its diversity and complexity and a measure of its mastery of natural forces, which can produce a widely varied built environment to serve the diverse needs of society.

Modern Building Practices

The Economic Context of Building Construction

Buildings, like all economic products, command a range of unit prices based on their cost of production and their value to the consumer. In aggregate, the total annual value of building construction in the various national economies is substantial. In 1987 in the United States, for example, it was about 10 percent of the gross domestic product, a proportion that is roughly applicable for the world economy as a whole. In spite of these large aggregate values, the unit cost of buildings is quite low when compared to other products. In the United States in 1987, new building cost ranged from about $0.50 to $2.50 per pound. The lowest costs are for simple pre-engineered metal buildings, and the highest represent functionally complex buildings with many mechanical and electrical services, such as hospitals and laboratories. These unit costs are at the low end of the scale of manufactures, ranking with inexpensive foodstuffs, and are lower than those of most other familiar consumer products. This scale of cost is a rough index of the value or utility of the commodity to society. Food, although essential, is relatively easy to produce; aircraft, at the high end of the scale, perform a desirable function but do so with complex and expensive mechanisms that command much higher unit prices which reflect not only the materials and labour required to produce them but also substantial capital and research investments. Buildings fall nearer to food in value; they are ubiquitous and essential, yet the services consumers expect them to provide can be supplied with relatively unsophisticated technology and inexpensive materials. Thus there has been a tendency for building construction to remain in the realm of low technology, for there has been relatively little incentive to invest in research given consumer expectations.

Within this general economic context, there are a number of specific parameters that affect the cost of buildings. First are government building codes, which are enacted to protect public health and safety; these take the form of both prescriptive and performance requirements. Structural requirements include description of the loads buildings must support, beginning with the constant everyday loads of building contents imposed by gravity and extending to the less frequent but more extreme loadings of wind and earthquake forces. These are specified on a statistical basis, usually the maximum expected to occur with a 100-year frequency. Safety factors for materials are specified to allow for accidental overloading and lapses of quality control. Economic considerations are also reflected; for example, buildings must perform well under normal gravity loads, but no code requires a building to resist direct exposure to the wind and low-pressure effects of a tornado, for its cost would be prohibitive.

Planning and zoning requirements provide for height and floor area limitations and building setbacks from lot lines to ensure adequate light and air to adjoining properties. Zoning regulations also establish requirements for permitted building usages, parking spaces, and landscaping and even set standards for the visual appearance of buildings. Another example is requirements for building atmosphere conditions; these include minimum (but not maximum) temperatures and rates of air change to dilute odours and provide an adequate oxygen supply. Life-safety requirements include adequate stairways for emergency exits, emergency lighting, smoke detection and control systems, and fire-resistant building materials. Sanitation requirements include adequate numbers of plumbing fixtures and proper pipe sizes. Electrical requirements include wire sizes, construction requirements for safety, and location of outlets.

Beyond the government standards there are market standards, which reflect user expectations for buildings. One example is elevator systems; elevators are not required by building codes, but in the United States, for example, the number of elevators in office buildings is calculated based on a maximum waiting period of 30 seconds. Cooling of building atmospheres is also not required by code but is provided in climates and building types where the marketplace has shown it to be cost-effective.

Building systems and components are perceived as having two dimensions of value. One is the purely functional dimension: the structure is expected to resist loads, the roof must keep out rain. The other is the aesthetic or psychic dimension: stone is perceived as more durable than wood; an elevator system with a waiting time of 30 seconds is preferable to one with a waiting time of two minutes. For these perceived differences many users are willing to pay more. When symbolic buildings such as temples, cathedrals, and palaces play an important role in society, the aesthetic dimension is important in valuing buildings; for example, the Parthenon of Athens or Chartres Cathedral commanded a level of investment in their economies that might be roughly compared to the U.S. Apollo space program. But in most buildings the functional dimension of value is dominant.

Because of its relatively low level of technology, wide geographic distribution, highly variable demand, and wide variety of building products, the building industry in industrialized countries is subdivided into many small enterprises. This lack of centralization tends to discourage research and keeps building components sturdy and simple, following well-tried formulas. Within this diversity there are a number of fairly well-defined markets based on building types; these include low-rise residential buildings, low-rise commercial, institutional, and industrial buildings, high-rise buildings, and long-span buildings.

A somewhat similar pattern is found in Eastern Europe, although the building industry there is more centralized. There is also a much smaller low-rise residential market, with most new housing being provided in high-rise buildings.

In developing countries the major market is for low-rise residential buildings to house rapidly growing populations. Much of the construction is undertaken by local craftsmen using simple building products. Local timber is widely used, and masonry materials still include the ancient mud brick. More sophisticated long-span and high-rise technologies are found only in major cities.

Building Design and Construction

Design Programming

The design of a building begins with its future user or owner, who has in mind a perceived need for the structure, as well as a specific site and a general idea of its projected cost. The user, or client, brings these facts to a team of design professionals composed of architects and engineers, who can develop from them a set of construction documents that define the proposed building exactly and from which it can be constructed.

Building design professionals include those licensed by the state—such as architects and structural, mechanical, and electrical engineers—who must formally certify that the building they design will conform to all governmental codes and regulations. Architects are the primary design professionals; they orchestrate and direct the work of engineers, as well as many other consultants in such specialized areas as lighting, acoustics, and vertical transportation.

The design professionals draw upon a number of sources in preparing their design. The most fundamental of these is building science, which has been gradually built up over the past 300 years. This includes the parts of physical theory that relate to building, such as the elastic theory of structures and theories of light, electricity, and fluid flow. There is a large compendium of information on the specific properties of building materials that can be applied in mathematical models to reliably project building performance. There is also a large body of data on criteria for human comfort in such matters as thermal environment, lighting levels, and sound levels that influence building design.

In addition to general knowledge of building science, the design team collects specific data related to the proposed building site. These include topographic and boundary surveys, investigations of subsoil conditions for foundation and water-exclusion design, and climate data and other local elements.

Concurrently with the collection of the site data, the design team works with the client to better define the often vague notions of building function into more precise and concrete terms. These definitions are summarized in a building space program, which gives a detailed written description of each required space in terms of floor area, equipment, and functional performance criteria. This document forms an agreement between the client and the design team as to expected building size and performance.

Design Development

The process by which building science, site data, and the building space program are used by the design team is the art of building design. It is a complex process involving the selection of standard building systems, and their adaptation and integration, to produce a building that meets the client's needs within the limitations of government regulations and market standards. These systems have become divided into a number of clear sectors by the building type for which they are intended. The design process involves the selection of systems for foundations, structure, atmosphere, enclosure, space division, electrical distribution, water supply and drainage, and other building functions. These systems are made from a limited range of manufactured components but permit a wide range of variation in the final product. Once the systems and components have been selected, the design team prepares a set of contract documents, consisting of a written text and conventionalized drawings, to describe completely the desired building configuration in terms of the specified building systems and their expected performance. When the contract documents have been completed, the final costs of the building can usually be accurately estimated and the construction process can begin.

Construction

Construction of a building is usually executed by a specialized construction team; it is normally separate from the design team, although some large organizations may combine both functions. The construction team is headed by a coordinating organization, often called a general contractor, which takes the primary responsibility for executing the building and signs a contract to do so with the building user. The cost of the contract is usually an agreed lump sum, although cost-plus-fee contracts are sometimes used on large projects for which construction begins before the contract documents are complete and the building scope is not fully defined. The general contractor may do some of the actual work on the building in addition to its coordinating role; the remainder of the work is done by a group of specialty subcontractors who are under contract to the general contractor. Each subcontractor provides and installs one or more of the building systems—e.g., the structural or electrical system. The subcontractors in turn buy the system components from the manufacturers. During the construction process the design team continues to act as the owner's representative, making sure that the executed building conforms to the contract documents and that the systems and components meet the specified standards of quality and performance.

Low-Rise Residential Buildings

Low-rise residential buildings include the smallest buildings produced in large quantities. Single-family detached houses, for example, are in the walk-up range of one to three stories and typically meet their users' needs with about 90 to 180 square metres (about 1,000 to 2,000 square feet) of enclosed floor space. Other examples include the urban row house and walk-up apartment buildings. Typically these forms have relatively low unit costs because of the limited purchasing power of their owners. The demand for this type of housing has a wide geographic distribution, and therefore most are built by small local contractors using relatively few large machines (mostly for earth moving) and large amounts of manual labour at the building site. The demand for these buildings can have large local variations from year to year, and small builders can absorb these economic swings better than large organizations. The building systems developed for this market reflect its emphasis on manual labour and its low unit costs. A proportion of single-family detached houses are "factory-built"; that is, large pieces of the building are prefabricated and then transported to the site, where considerable additional work is required to complete the finished product.

Foundations

All foundations must transmit the building loads to a stable stratum of earth. There are two criteria for stability: first, the soil under the foundations should be able to receive the imposed load without more than about 2.5 centimetres (one inch) of settlement and, second, the settlement should be uniform under the entire building. It is also important that the bottom of the foundation be below the maximum winter frost level. Wet soil expands as it freezes, and repeated freeze–thaw cycles can move the building up and down, leading to possible displacement and damage. Maximum frost depth varies with climate and topography. It can be as deep as 1.5 metres (five feet) in cold continental climates and is zero in tropical and some subtropical areas. The foundation systems for low-rise residential buildings are suitable for their light loads; nearly all are supported on spread footings, which are of two types—continuous footings that support walls and isolated pad footings that support concentrated loads. The footings themselves are usually made of concrete poured directly on undisturbed soil to a minimum depth of about 30 centimetres (12 inches). If typical continuous concrete footings are used, they usually support a foundation wall that acts either as a retaining wall to form a basement or as a frost wall with earth on both sides. Foundation walls can be built of reinforced concrete or masonry, particularly concrete block. Concrete blocks are of a standard size larger than bricks and are hollow, forming a grid of vertical planes. They are the least expensive form of masonry—using cheap but strong material—and their large size economizes on the labour required to lay them. Their appearance and weathering properties are inferior to those of fired masonry, but they are satisfactory for foundation walls. In some places timber foundation walls and spread footings are used. Excavation for foundations is the most highly mechanized operation in this building type; it is done almost entirely with bulldozers and backhoes.

Structural Systems

Timber Frames

In these small buildings the ancient materials of timber and masonry are still predominant in the structural systems. In North America, which has abundant softwood forests, light timber frames descended from the 19th-century balloon frame are widely used. These present-day "platform" frames are made of standard-dimension timbers, usually two or four centimetres (0.75 or 1.5 inch) thick, which are joined together by machine-made nails and other metal fasteners using hand tools.

Timber frame construction: The completed frame of a modern timber-frame house.

The first step is to construct a floor, which rests on the foundation wall. A heavy timber sill is attached to the wall with anchor bolts, and on top of it are nailed the floor joists, typically 4 × 28 centimetres (1.5 × 11.25 inches) and spaced 40 centimetres (16 inches) apart. The span of the floor joists is usually about 3.6 metres (12 feet), which is the common maximum length of available timbers. The floor may need intermediate supports in the form of interior foundation walls or, if there is a basement, intermediate beams of wood or steel supported by the foundation walls and columns. For longer spans, floor trusses can be made, with members joined by nail grids or nailed plywood gussets or with wood chords and diagonal metal web members. On top of the joists is nailed plywood subflooring, which forms the deck and gives lateral stability to the floor plane.

The exterior bearing walls are made of 4 × 9-centimetre (1.5 × 3.5-inch; "2 × 4") timber verticals, or studs, spaced 40 or 60 centimetres (16 or 24 inches) apart, which rest on a horizontal timber, or plate, nailed to the floor platform and support a double plate at the top. The walls are sheathed on the outside with panels of plywood or particleboard to provide a surface to attach the exterior cladding and for lateral stability against wind. Plywood and particleboard are fabricated in panels of standard sizes. Plywood is made of thin layers of wood, rotary-cut from logs and glued together with the wood grain running perpendicularly in adjoining layers. Particleboard consists of fine wood chips mixed together in an adhesive matrix and allowed to harden under pressure. On top of the wall plate is placed either a second floor or the roof.

Since most of the roofing materials used in these buildings are not fully watertight, the roofs must have sloped surfaces to rapidly drain off rainwater. Sloped forms are created by two methods. The traditional method uses joists similar to those of floor construction to span between exterior walls. Rafters are nailed to the ends of each joist and the rafters meet at a central ridge member, forming a triangular attic space. Where no attic space is needed, it has proved more economical to span the roof with triangular trusses with interior web members. These roof trusses are usually made of narrow timbers joined by nails, glue, or metal connectors and they are often prefabricated in a workshop. Plywood or particleboard sheathing is then nailed to the roof surfaces to receive the roofing and to provide lateral stability, making the entire frame into a rigid box.

Light timber frames are quite flammable, but small one- or two-story buildings are easy to evacuate in case of a fire, and building codes permit the use of these frames with such features as fire-resistant gypsum board on the interiors and fire-stops (short wooden members) between the studs. Timber structures are attacked by certain species of insects—such as termites and carpenter ants—as well as certain fungi, particularly in warm, moist climates. Wood can be chemically treated to discourage these attacks; other precautions include raising the timber above the ground and keeping it dry.

Masonry Walls

Structural masonry walls are also used in this building type, primarily in multistory buildings, where they offer greater load-bearing capacity and fire resistance. Brick and concrete block are the major materials, brick being favoured for exterior surfaces because of its appearance and durability. Solid brick walls are rarely used, due to the higher labour and material costs; composite walls of brick and block or block alone are common. Cavity walls are used in colder climates; in these, two wythes (vertical layers) of masonry are built on either side of a layer of rigid insulation. The wythes are joined together by steel reinforcement that runs through the insulation and is laid in the horizontal masonry joints at intervals. Cavity walls have a heat-flow rate that is 50 percent of that of a solid wall. Timber floor and roof construction, similar to balloon framing, is used with masonry construction; and there is also some use of precast prestressed hollow concrete panels, which are fireproof and can span up to nine metres (30 feet).

Enclosure Systems

Enclosure systems for this building type are varied. For roofs, traditional wood shingles or, more commonly, felt asphalt shingles are used, as are semicylindrical clay tiles and standing-seam metal roofs. Rainwater from roofs is usually caught in metal gutters and directed to exterior downspouts that discharge onto splash blocks or into underground drains connected to storm sewers.

The wall surfaces of low-rise residential buildings are clad with a range of different materials. Traditional wood elements such as shingles and horizontal shiplap, or clapboard siding, are used on light timber frames as are vertical tongue-and-groove siding and boards and battens. Aluminum and vinyl sidings have been adapted from these wooden forms. Brick and stone veneer are also applied over timber and anchored to it with metal fasteners. Cement plaster, or stucco, is another traditional material used to enclose both timber and masonry structures, and its semiliquid application allows great plasticity of form. A more recent development is a very thin synthetic resin stucco applied directly to the surface of rigid plastic foam insulation.

Insulation, which slows the rate of heat transfer through the enclosure, is usually applied at all exterior building surfaces that are exposed to air. There are two major types of insulation, rigid and nonrigid. Rigid insulations are primarily plastic foams (the dead air in the foam cells is the true insulator), which vary in thickness from 2.5 to five centimetres (one to two inches). They include styrofoam, used primarily below grade behind frost walls due to its low fire resistance; urethane foam; isocyanurate foam, which has the best fire resistance; and foam glass. Nonrigid insulations are usually made of fibre—glass fibre being the most common—often with a foil-backed paper on one side. Fibre insulations are made in thicknesses up to 23 centimetres (9.25 inches). The effectiveness of an insulation material is measured in terms of its heat-transfer rate, or U-value, often expressed as the number of BTUs passing through a given unit of insulating material each hour at an expressed temperature differential across the material. Low U-values indicate good insulating properties of the material. U-value is an inverse function of thickness, so that there is a limit to the cost-effectiveness of increasing the amount of insulation on a surface. Rigid insulation panels are applied to vertical wall sheathing and the surfacing material is fastened through the insulation, or it is applied to horizontal roof decks. Glass fibre is usually applied in the spaces between wall studs and between roof joists or the bottom chords of roof trusses.

Most low-rise residential buildings have a limited number of transparent openings in their exteriors, because of the traditional requirements of interior privacy and the relatively higher cost of windows compared to opaque walls. The traditional wooden frames of domestic windows are often clad in extruded vinyl or aluminum cladding, and frames made entirely of extruded aluminum are common. Residential windows are a major means of ventilation, and there are a variety of operating actions for their movable sections: sliding or double-hung windows are still the major form, but hinged types—including casement, hopper, and awning forms—are also used. Sliding glass panel doors are also used, particularly in warmer regions. Glazing is still largely of clear glass. Double glazing, with two panes bonded to a metal tubular separator that contains a desiccant, is cost-effective in northern climates, but triple glazing is used commonly only in regions above about 55° to 60° latitude. A recent development is heat-mirror glass, in which a low-emissivity coating enhances the relative opacity of the glass to infrared radiation and slows the rate of internal heat loss in winter.

Interior Finishes

Interior finishes and space-division systems define the living spaces within residential buildings with a range of both natural and synthetic materials. The most widely used wall finish is gypsum board, a prefabricated form of traditional wet plaster. Wet gypsum plaster is cast between paper facings to form large panels that are nailed to light timber or metal frameworks. The joints between the panels are filled with a hard-setting resin compound, giving a smooth seamless surface that has considerable fire resistance. Gypsum board forms the substrate to which a number of other materials, including thin wood-veneered plywood and vinyl fabrics, can be applied with adhesives. In wet areas such as kitchens and bathrooms, water-resistant gypsum board is used, sometimes with the addition of adhesive-applied ceramic tile.

Doors in residential buildings are usually of the hollow-core type, with thin veneers of wood glued over a honeycomb paper core and solid wood edge strips; door frames are typically made of machined timber shapes. Plastic laminates bonded to particleboard are extensively used for built-in cabinets and countertops. The most common floor finish is carpeting, most of which is now made of synthetic fibres, displacing the traditional wool and cotton. It can be easily maintained, and its soft visual and tactile texture, as well as its sound-absorbing qualities, make it attractive for residential use. Hardwoods—primarily oak, birch, and maple—are also used for floors, both in the traditional narrow planks nailed to plywood decks and as prefabricated parquet elements, which are applied with adhesives. In wet or hard-use areas vinyl-composition tiles or ceramic tiles are used.

Plumbing

Domestic water-supply systems for low-rise residential buildings have two sources, either municipal water-distribution systems or, where these are not available, wells that are drilled to underground aquifers which are free of contamination. Water is drawn from the wells with small submersible electric pumps, which are lowered through the well casing to the intake. Underground exterior water-supply pipes are usually cast-iron with threaded connections to contain the pressures applied to the fluid, which is typically sufficient to raise it four stories. Within the building, copper tubing with soldered connections is used for distribution because of its corrosion resistance and ease of fabrication; in some areas plastic pipe is also used. The domestic water supply is divided into cold and hot systems, the cold water being piped directly to the fixtures. The hot-water system first draws the supply through a hot-water heating tank, which raises its temperature to about 60 °C (140 °F) using electric resistance or gas heat. Domestic water heaters that use solar radiation to heat water in coils exposed to the sun on a glass-covered black metal plate (flat-plate solar collectors) are found in areas where there is ample sunshine and relatively high energy costs. The hot water is then distributed from the heater to the fixtures in a recirculating loop pipe system, in which gravity and temperature differentials maintain a constant temperature in period of low demand.

The primary residential use of water is in the bathroom, which typically includes a bathtub of cast iron or pressed steel with a ceramic porcelain coating (although fibre-glass-reinforced resin is also used), a ceramic lavatory, and a ceramic tank-type water closet. The bath and lavatory are supplied with hot and cold water through faucets with lever or screw-type valve controls. The valve of the water closet supply is also lever-operated and relies on the gravity power of the water in the tank for its flushing action. Shower baths are also common, often incorporated into bathtub recesses or in a separate compartment finished with ceramic tile. In some countries a bidet is included.

Other widely used plumbing fixtures include kitchen sinks, usually of cast iron or pressed steel with a ceramic porcelain coating or of stainless steel; automatic dishwashing machines; and automatic washing machines for laundry. Kitchen sinks can be fitted with garbage disposals, which grind solid waste into a fluid slurry that is flushed out with wastewater. Where the possibility of back siphonage of wastewater into the water supply exists, a vacuum breaker must be provided at the supply to prevent this happening, but most domestic plumbing fixtures are designed to avoid this possibility.

Drainage systems to remove wastewater are made of cast-iron pipe with threaded joints or bell-and-spigot joints sealed with molten lead or with plastic pipe with solvent-welded joints. The waste pipe of every plumbing fixture is provided with a semicircular reverse curve, or trap, which remains constantly filled with water and prevents odours from the drainage system from escaping into occupied spaces. Immediately downstream from each trap is an opening to a vent pipe system, which lets air into the drainage system and protects the water seals in the traps from removal by siphonage or back pressure. When wastewater leaves the building, it is drained through a back-flow-prevention valve and into underground ceramic pipes. It then flows by gravity to either a private sewage treatment plant, such as septic tank and tile field, or to the public sewer system. If the discharge level of the wastewater is below the level of the sewer, a sewage ejector pump is required to raise the wastewater to a higher level, where gravity carries it away.

Heating and Cooling

Atmosphere-control systems in low-rise residential buildings use natural gas, fuel oil, or electric resistance coils as central heat sources; usually the heat generated is distributed to the occupied spaces by a fluid medium, either air or water. Electric resistance coils are also used to heat living spaces directly with radiant energy. Forced-air distribution moves the heat-bearing air through a treelike system of galvanized sheet-metal ducts of round or rectangular cross section; electric-powered fans provide a pressure differential to push the air from the heat source (or furnace) to the living spaces, where it is expelled from grills located in the walls or floors. The negative pressure side of the fan is connected to another treelike system of return air ducts that extract air from living spaces through grills and bring it back to the furnace for reheating. Fresh outside air can be introduced into the system airstream from an exterior intake, and odour-laden interior air can be expelled through a vent, providing ventilation, usually at the rate of about one complete air change per hour. To conserve energy, air-to-air heat exchangers can be used in the exhaust–intake process. The heated air is usually supplied in constant volume, and the ambient temperature is varied in response to a thermostat located in one room. Central humidity control is rarely provided in this building type.

Another common heating system is the radiant hot-water type. The heat source is applied to a small boiler, in which water is heated and from which it is circulated by an electric pump in insulated copper pipes similar to a domestic hot-water system. The pipes can be connected to cast-iron or finned tube steel radiators within the living spaces. The radiators are placed near the areas of greatest heat loss (such as windows or outside walls) where their radiant energy heats the surrounding air and creates a convection cycle within the room, producing a roughly uniform temperature within it. The hot water can also be conducted through narrow pipes placed in a continuous looping pattern to create a large radiant surface; this pattern of pipes may be cast in a concrete floor slab or placed above a ceiling to heat the adjoining living space. Temperature control in hot-water systems uses a thermostat in the living space to adjust the pumped flow rate of the water to vary the heat supplied.

Radiant electric resistance heating systems use coils in baseboard units in the rooms, which create convection cycles similar to hot-water radiators, or resistance cables in continuous looped patterns embedded in plaster ceilings. Local temperature control can be much more precise with electric heating, because it is possible to install a thermostatically controlled rheostat to vary the energy output of relatively small sections of baseboard units or cable.

A type of space heating that is increasing in use in residential buildings is passive solar radiation. On sunny winter days, south-facing windows let in substantial amounts of energy, often enough to heat the entire building. Wood-burning fireplaces with chimneys are still widely provided in residential buildings, but their use is mostly for aesthetic effect.

The cooling of atmospheres in low-rise residential buildings is often done locally with unit air conditioners, which penetrate the exterior wall of the space to be cooled; this permits the intake of fresh air when desired and the ejection of heat pumped from the space to the exterior air. Less often, forced-air heating systems have cooling coils introduced into the airstream to provide a centrally cooled interior. A compressive cooling process is used, similar to that in a domestic refrigerator. A refrigerant, which is a liquid at room temperature, is pumped through a closed system of coiled copper tubes. An electric pump maintains a low pressure in the cooling coils, and the liquid refrigerant passes through an expansion valve from a region of high pressure to the low-pressure coils. This change in pressure results in a phase change of the refrigerant; it turns from a liquid into a gas and absorbs heat in the process, just as water absorbs heat when it is boiled and converted into steam. The heat absorption of the liquid-to-gas transition cools air passing over the cooling coils. The cooled air is circulated through the building by the forced-air system. When the low-pressure gaseous refrigerant leaves the cooling coils, it goes through the pump and is pressurized. The refrigerant travels through condensing coils, which are located outside the building; there the phase change is reversed as the gas turns to a high-pressure liquid and liberates heat to the exterior air passing over the condensing coils. The liquid refrigerant returns to the expansion valve to repeat the cooling cycle. The refrigeration machine is thus a "heat pump" that moves heat out of the building to the exterior atmosphere. Heat pumps can also be run in reverse in the winter months to pump heat from the outside air into the building interior; they work best in mild climates with fairly warm winter temperatures. The use of heat pumps in cold climates poses many difficult technological problems.

Interior atmospheres are also ventilated by operating windows, as well as by unintended leakage at all types of exterior openings. Bathrooms, kitchens, and laundries generate odours and heat and often have separate exhaust systems powered by electric fans that are operated intermittently as required. Residential atmosphere quality is also protected by the smoke detector, which sounds an alarm to warn of possible danger when smoke reaches even a very low level in living spaces.

Electrical Systems

Electrical systems in residential buildings are supplied from public utility power grids, starting from a step-down transformer near the building that reduces the high line voltage to a safer level. An underground or overhead cable from the transformer leads to the building, where it is connected to a meter that records the energy used by the subscriber. Immediately beyond the meter is a fused main switch to protect the building against an accidental power surge from the grid. The main service is then broken down into a number of circuits by a panelboard, each circuit having a fused switch. From the panelboard the wires of each circuit distribute the electricity to different areas of the building. The wires are usually copper, although aluminum is also used, and are covered with thermoplastic insulation. The wires must be contained in conduit, which is either metal or plastic tubing, to protect against damage and reduce the possibility of fire in the case of accidental overloading of the wires. Conduits are usually concealed in finished spaces within the framing of partition walls or above ceilings and terminate in junction boxes flush with a wall surface. The junction boxes contain terminal devices such as the convenience outlet, control switches, or the connection point for built-in light fixtures.

Residential lighting is provided primarily by movable incandescent fixtures plugged into convenience outlets, but there is often built-in lighting in kitchens, bathrooms, corridors, and closets, mostly of the incandescent type. There is also some use of fluorescent lighting, particularly in built-in fixtures. Overall interior light levels in residential uses are low, about 20–40 footcandles. Exterior lighting is used for entrances, walkways, and exterior living spaces.

The power densities of dwelling units are fairly low and are declining because of the increased use of fluorescent lighting fixtures and improvement of efficiency in electrical appliances. The decline in power consumption enhances the prospect of the widespread appearance of dwellings—particularly detached houses—with their own independent electric power generation and storage systems, unconnected to public utility grids. Photovoltaic cells, which convert sunlight directly into electricity, in combination with storage batteries can offer these residences a new kind of energy autonomy.

Low-Rise Commercial, Institutional and Industrial Buildings

The size of buildings in the commercial, institutional, and industrial market segment ranges from a few hundred to as much as 45,000 square metres (500,000 square feet). All of these buildings have public access and exit requirements, although their populations may differ considerably in density. The unit costs are generally higher than those for dwellings (although those of simple industrial buildings may be lower), and this type includes buildings with the highest unit cost, such as hospitals and laboratories. Residential buildings are fairly static in their function, changing only at long intervals. By contrast, most commercial, institutional, and industrial buildings must respond to fairly rapid changes in their functions, and a degree of flexibility is required in their component systems. In addition, these buildings are built by contractors who utilize heavy mechanized equipment not only for foundations (pile drivers and caisson augers) but also for lifting heavy components (a wide variety of cranes and hoists). Semimanual machines such as cement finishers, terrazzo grinders, and welding generators are also used, but a large percentage of the work is done manually; the human hand and back remain major instruments of the construction industry, well adapted to the nonrepetitive character of building.

Foundations

The foundations in these buildings support considerably heavier loads than those of residential buildings. Floor loadings range from 450 to 1,500 kilograms per square metre (100 to 300 pounds per square foot), and the full range of foundation types is used for them. Spread footings are used, as are pile foundations, which are of two types, bearing and friction. A bearing pile is a device to transmit the load of the building through a layer of soil too weak to take the load to a stronger layer of soil some distance underground; the pile acts as a column to carry the load down to the bearing stratum. Solid bearing piles were originally made of timber, which is rare today; more commonly they are made of precast concrete, and sometimes steel H-piles are used. The pile length may be a maximum of about 60 metres (200 feet) but is usually much less. The piles are put in place by driving them into the ground with large mechanical hammers. Hollow steel pipes are also driven, and the interiors are excavated and filled with concrete to form bearing piles; sometimes the pipe is withdrawn as the concrete is poured. An alternative to the bearing pile is the caisson. A round hole is due to a bearing stratum with a drilling machine and temporarily supported by a steel cylindrical shell. The hole is then filled with concrete poured around a cage of reinforcing bars; and the steel shell may or may not be left in place, depending on the surrounding soil. The diameter of caissons varies from one to three metres (three to 10 feet). The friction pile of wood or concrete is driven into soft soil where there is no harder stratum for bearing beneath the site. The building load is supported by the surface friction between the pile and the soil.

When the soil is so soft that even friction piles will not support the building load, the final option is the use of a floating foundation, making the building like a boat that obeys Archimedes' principle—it is buoyed up by the weight of the earth displaced in creating the foundation. Floating foundations consist of flat reinforced concrete slabs or mats or of reinforced concrete tubs with walls turned up around the edge of the mat to create a larger volume.

If these buildings do not have basements, in cold climates insulated concrete or masonry frost walls are placed under all exterior nonbearing walls to keep frost from under the floor slabs. Reinforced concrete foundation walls for basements must be carefully braced to resist lateral earth pressures. These walls may be built in excavations, poured into wooden forms. Sometimes a wall is created by driving interlocking steel sheet piling into the ground, excavating on the basement side, and pouring a concrete wall against it. Deeper foundation walls can also be built by the slurry wall method, in which a linear series of closely spaced caissonlike holes are successively drilled, filled with concrete, and allowed to harden; the spaces between are excavated by special clamshell buckets and also filled with concrete. During the excavation and drilling operations, the holes are filled with a high-density liquid slurry, which braces the excavation against collapse but still permits extraction of excavated material. Finally, the basement is dug adjoining the wall, and the wall is braced against earth pressure.

Timber

The structures of these buildings are mostly skeleton frames of various types, because of the larger spans their users require and the need for future flexibility. Timber is used, but on a much-reduced scale compared to residential buildings and primarily in regions where timber is readily available. The public nature of commercial and institutional buildings and the hazards of industrial buildings generally require that they be of noncombustible construction, and this largely excludes the use of light timber frames. Heavy timber construction can be used where the least dimensions of the members exceed 14 centimetres (5.5 inches); when timbers are this large they are charred but not consumed in a fire and are considered fire-resistant. Because most harvested trees are fairly small, it is difficult to obtain solid heavy timbers, and most large shapes are made up by glue laminating smaller pieces. The synthetic glues used are stronger than the wood, and members with cross sections up to 30 × 180 centimetres (12 × 72 inches) are made; these may be tapered or otherwise shaped along their length. Skeletons of glue-laminated beams and columns, joined by metal connectors, can span 30 to 35 metres (100 to 115 feet). Heavy decking made of tongue-and-groove planks up to 9.4 centimetres (3.75 inches) thick is used to span between beams to support floors and roofs.

Steel

Steel is a major structural material in these buildings. It is a strong and stiff material and yet relatively inexpensive, and it can be quickly fabricated and erected, which saves construction time. Although steel is noncombustible, it starts to lose strength when heated above 400 °C (750 °F), and building codes require it to be fireproofed in most multistory buildings; in small and low-hazard buildings, however, it can be left unprotected.

Nearly all structural steel—including sheets, round or square bars, tubes, angles, channels, and I beam or wide flange shapes—is formed by the hot-rolling process. Steel roof and floor deck panels are fabricated from sheet metal by further cold-rolling into corrugated profiles four to eight centimetres (1.5 to three inches) deep and 60 centimetres (24 inches) wide. They are usually welded to the supporting steel members and can span up to 4.5 metres (15 feet). The lightest and most efficient structural shape is the bar (or open web) joist, a standard truss made with angles for the top and bottom chords, joined by welding to a web made of a continuous bent rod. It is used almost exclusively to support roofs and can span up to 45 metres (150 feet). The standard rolled shapes are frequently used as beams and columns, the wide flange, or W shape, being the most common. The widely separated flanges give it the best profile for resisting the bending action of beams or the buckling action of columns. W shapes are made in various depths and can span up to 30 metres (100 feet). Where steel beams support concrete floor slabs poured onto a metal deck, they can be made to act compositely with the concrete, resulting in considerable economies in the beam sizes.

The connections of steel shapes are of two types: those made in the workshop and those made at the building site. Shop connections are usually welded, and site or field connections are usually made with bolts due to the greater labour costs and difficulties of quality control in field welding. Steel columns are joined to foundations with base plates welded to the columns and held by anchor bolts embedded in the concrete. The erection of steel frames at the building site can proceed very rapidly, because all the pieces can be handled by cranes and all the bolted connections made swiftly by workers with hand-held wrenches.

A large proportion of steel structures are built as prefabricated, pre-engineered metal buildings, which are usually for one-story industrial and commercial uses. They are manufactured by companies that specialize in making such buildings of standard steel components—usually rigid steel bents or light trusses—which are assembled into frames and enclosed with corrugated metal siding. The configurations can be adapted to the needs of individual users. The metal building industry is a rare example of a successful application of prefabrication techniques in the construction industry in the United States, where its products are ubiquitous in the suburban and rural landscape.

Concrete

Reinforced concrete is also a major structural material in these buildings. Indeed, outside of North America and Western Europe, it is the dominant industrialized building material. Its component parts are readily available throughout the world at fairly low cost. Portland cement is easily manufactured by burning shale and limestone; aggregates such as sand and crushed limestone can be easily obtained. Steel minimills, which use scrap iron to feed their electric furnaces, can mass-produce reinforcing bars for regional use. In industrialized countries the mixing and delivery of liquid concrete to building sites has been mechanized with the use of central plants and mixing trucks, and this has substantially reduced its cost. In barely 100 years, reinforced concrete has risen from an experimental material to the most widespread form of building construction.

There are two methods of fabricating reinforced concrete. The first is to pour the liquid material into forms at the building site; this is so-called in situ concrete. The other method is called precast concrete, in which building components are manufactured in a central plant and later brought to the building site for assembly. The components of concrete are portland cement, coarse aggregates such as crushed stone, fine aggregates such as sand, and water. In the mix, water combines chemically with the cement to form a gel structure that bonds the stone aggregates together. In proportioning the mix, the aggregates are graded in size so the cement matrix that joins them together is minimized. The upper limit of concrete strength is set by that of the stone used in the aggregate. The bonding gel structure forms slowly, and the design strength is usually taken as that occurring 28 days after the initial setting of the mix. Thus there is a one-month lag between the time in situ concrete is poured and the time it can carry loads, which can significantly affect construction schedules.

In situ concrete is used for foundations and for structural skeleton frames. In low-rise buildings, where vertical gravity loads are the main concern, a number of framing systems are used to channel the flow of load through the floors to the columns for spans of six to 12 metres (20 to 40 feet). The oldest is the beam and girder system, whose form was derived from wood and steel construction: slabs rest on beams, beams rest on girders, and girders rest on columns in a regular pattern. This system needs much handmade timber formwork, and in economies where labour is expensive other systems are employed. One is the pan joist system, a standardized beam and girder system of constant depth formed with prefabricated sheet-metal forms. A two-way version of pan joists, called the waffle slab, uses prefabricated hollow sheet-metal domes to create a grid pattern of voids in a solid floor slab, saving material without reducing the slab's strength. The simplest and most economical floor system is the flat plate, where a plain floor slab about 20 centimetres (eight inches) thick rests on columns spaced up to 6.7 metres (22 feet) apart. If the span is larger, the increasing load requires a local thickening of the slab around the columns. When these systems are applied to spans larger than nine to 12 metres (30 to 40 feet), a technique called posttensioning is often used. The steel reinforcing takes the form of wire cables, which are contained in flexible tubes cast into the concrete. After the concrete has set and gained its full strength, the wires are permanently stretched taut using small hydraulic jacks and fastening devices, bending the entire floor into a slight upward arch. This reduces deflection, or sagging, and cracking of the concrete when the service load is applied and permits the use of somewhat shallower floor members. Concrete columns are usually of rectangular or circular profile and are cast in plywood or metal forms. The reinforcing steel never exceeds 8 percent of the cross-sectional area to guard against catastrophic brittle failure in case of accidental overloading.

Precast concrete structural members are fabricated under controlled conditions in a factory. Members that span floors and roofs are usually pretensioned, another prestressing technique, which is similar in principle to posttensioning. The reinforcement is again steel wire, but the wires are put into tension (stretched) on a fixed frame, formwork is erected around the taut wires, and concrete is poured into it. After the concrete has set and gained its full strength, the wires are cut loose from the frame. As in posttensioning, this gives the precast floor members a slight upward arch, which reduces deflection and permits the use of shallower members. Precast prestressed floor elements are made in a number of configurations. These include beams of rectangular cross section, hollow floor slabs 15 to 30 centimetres (six to 12 inches) deep and spanning up to 18 metres (60 feet), and single- and double-stem T shapes up to 1.8 metres (six feet) deep and spanning up to 45 metres (150 feet). Precast concrete columns are usually not prestressed and have projecting shelves to receive floor members. At the building site, precast members are joined together by a number of methods, including welding together metal connectors cast into them or pouring a layer of in situ concrete on top of floor members, bonding them together. Precast prestressed construction is widely used, and it is the dominant form of construction in the Soviet Union and Eastern Europe.

Masonry finds only a limited structural use in these buildings. Concrete block walls with brick facing and punched openings (discrete windows entirely surrounded by the facing material) spanned by concealed steel lintels can be used for exterior bearing walls where the interior is a skeleton frame of steel or timber. The use of interior bearing walls so greatly reduces the flexibility needed in these buildings that they are only rarely found.

Enclosure Systems

Enclosure systems in these buildings range from rather simple forms in industrial uses to quite sophisticated assemblies in the commercial and institutional sectors. Most have in common the use of flat roofs with highly water-resistant coverings, the traditional one being a built-up membrane of at least four layers of coal-tar pitch and felt, often weighted down with a gravel ballast. Such roofs are pitched at slopes of 1 : 100 to 1 : 50 toward interior drains. In recent years the single-ply roof, made of plastic membranes of various chemistries, has found wide application. The seams between the pieces of membrane are heat- or solvent-welded together, and they are either ballasted with gravel or mechanically fastened to the underlying substrate, which is usually rigid foam insulation. Sometimes standing-seam sheet-metal roofs are also used; the best quality is continuously welded stainless steel.

The choice of transparent surfaces in these enclosures is based on three major considerations: conductive heat transfer, radiant energy transfer, and safety. All the transparent materials used in the low-rise residential sector are found, plus a number of others. In buildings with fully controlled atmospheres, double glazing is common to reduce heat transfer and both interior and exterior condensation on the glass. Commercial and institutional buildings tend to have large internal sources of heat gain, such as people and lighting, so it is desirable to exclude at least some solar gain through the transparent surfaces to reduce energy consumption in cooling. This can be done by reducing the light transmission or shading coefficient of the glass by integrally tinting it in various colours; grey, bronze, and green are common tints. This can also be accomplished by vacuum-plating partial reflective coatings of varying densities to an inner surface of double glazing; this can reflect up to 90 percent of the incident energy. Two kinds of reflecting metal are used: aluminum, which is silver in tone, and rubidium, which is gold-toned. These coatings are perceived as strong tints when the outside world is viewed through them by day: grey for aluminum and green for rubidium.

Skylights or horizontal transparent surfaces have found wide application in these types of buildings. These installations range from purely functional daylighting in industrial uses to elaborate aesthetic forms in commercial structures. In horizontal applications, and in vertical walls where people might blunder into glazed panels, safety glazing is required. Safety glazing is of four types: certain plastics that are flexible and difficult to break; wire-embedded glass, which holds together when broken; tempered glass, which is very strong and breaks into tiny and relatively harmless fragments; and laminated glass, which consists of two layers of glass heat-welded together by an intermediate plastic film. Laminated glass can also be made with tinted lamination film, producing many colours not available in integrally coloured glass.

Because many of these buildings have skeleton structures, their vertical surfaces are enclosed in nonstructural curtain walls that resist wind forces and provide weatherproofing. Curtain walls are of several types; the most common is one supported by a metal (typically aluminum) gridwork attached to the building structure. The vertical members, called mullions, are attached to the building at every floor and are spaced 1.5 to three metres (five to 10 feet) apart; the horizontal members, called muntins, are attached between the mullions. The rectangles between the grid of mullions and muntins are filled with transparent or opaque panels. The transparent surfaces can be any of those just described, and the opaque panels include opaque coloured glass, painted or anodized aluminum sheets, porcelain enameled steel sheets, fibreglass-reinforced cement, and stone wafers of granite, marble, or limestone cut with diamond-edged tools. All of these materials are usually backed up by rigid insulation to slow heat transfer. Metal sandwich panels are also used for economy of material; two thin layers of metal are separated by a core of different material, often with a high U-value for insulating effect. The separation of the thin layers of strong metal greatly increases the overall stiffness of the panel. The joints between panels and the supporting grid are weatherproofed with elastomeric sealants (cold-setting synthetic rubbers) or by prefabricated rubber gaskets. In glazed areas of curtain walls, mullions of structural glass are an alternative to metal mullions; they are more expensive, but they give an effect of greater transparency where this is desired.

Another type of curtain wall is the panel type. It has no gridwork of mullions and muntins but is made of large prefabricated rigid panels connected to the floors and spanning between them, with transparent openings made as holes cut out of the panel. The panels can be made of precast concrete, aluminum, or steel, often in sandwich form; elastomeric sealants are used to close the joints.

The finishes of metals in curtain walls include anodizing of aluminum, an electrolytic process that builds up the natural colourless oxide of aluminum into a thick adherent layer; it often includes the introduction of colour into the oxide layer itself. Durable paint coatings (with lifetimes of up to 40 years) can be applied to the metal in the factory; more conventional paints that must be renewed at shorter intervals are also used.

Interior Finishes

Partitions

Space-division systems in these buildings make use of gypsum board partitions, usually applied to a framework of formed sheet-metal members attached to the building structure. They are readily demolished and rebuilt at relatively low cost, meeting the need for flexibility in such buildings. They are often used for fire-resistive protective enclosures, for which a number of layers are laminated to achieve the specified fire resistance. Transparent and translucent partitions are also used, with different types of glass set in metal frames. Office buildings may contain prefabricated movable metal partitions, which typically use metal sandwich panel construction to create panels with both transparent and opaque surfaces as well as doors. These partitions are expensive compared with gypsum board and must be moved often to justify the greater initial cost. Concrete block is used in unfinished spaces and for fire-resistive partitions. Glazed ceramic block or ceramic tile applied over concrete block or gypsum board is used in wet areas and where cleanliness is a problem, such as in kitchens and toilet rooms. Occasionally walls with wood paneling or stone veneer are used for aesthetic effect. Doors are usually set in formed sheet-metal frames, although some wood frames are used. The doors themselves are usually made of solid timbers glue-laminated together and covered with thin decorative wood veneers; painted hollow sheet-metal doors are used for exterior doors and in areas of hard use.

Ceiling Finishes

Ceiling finishes in these buildings create a sandwich space below the roof or floor slab above, which conceals projecting structural elements, recessed light fixtures, electrical wiring conduits, and air-handling ductwork. The ceiling must be accessible to change or maintain the service elements located above it, and the most common ceiling system is composed of wet felted mineral fibre panels, painted and perforated on one side for sound absorption. The removable panels are supported on a grid of formed sheet-metal tee bars or zee tracks, which are suspended by wires from the structure above. Where accessibility is not important and a smooth finish is desired, suspended gypsum board ceilings can be used.

Floor Finishes

Floor finishes in commercial and institutional uses make considerable use of synthetic-fibre carpeting and vinyl composition tile. In areas of higher traffic harder surfaces may be used—for example, cut stone tiles of marble or granite, ceramic tile applied with epoxy adhesive to the substrate, or terrazzo. Terrazzo is made in two ways, traditional and thin-set. In the traditional form a four-centimetre (1.5-inch) layer of cement and sand grout is poured over the substrate; a grid of metal divider strips to control shrinkage cracks is set on the hardened surface, and grout mix of coloured cement and marble chips is poured between the strips. After hardening, the surface is machine polished to expose the marble chips and metal dividers. Thin-set terrazzo is made by placing the metal strips and pouring the binder and marble chips directly onto the subfloor, without the underbed of cement and sand. It is generally possible only when epoxy resins are used in place of cement binders. Terrazzo is available in many colours, and it forms a hard, smooth, and durable surface that is easily cleaned.

Life-Safety Systems

Most important in the hierarchy of interior elements are life-safety systems to protect and evacuate the building population in emergencies. These include life-threatening events, such as fire and smoke and earthquakes, and less critical ones, such as electric power failures. To deal with the threat of fire and smoke there is an array of fire-detection and fire-suppression systems. These include electronic heat and smoke detectors that can activate audible alarm devices to warn the building population and automatically notify local fire departments. For fire suppression hand-operated fire extinguishers must be provided, but many buildings have a separate piping system to provide water for firefighting. If public water mains cannot provide adequate water pressure, an electric pump is included, and there is also a connection outside the building to attach portable fire truck pumps. The piping terminates in an array of sprinkler heads located throughout the building in the ceiling plane in a density ranging from eight to 18 square metres (90 to 200 square feet) per head. Typically there is always water in the pipes (a wet system), though dry systems are used in unheated buildings or where leakage might damage the contents. The head is opened to spray water by a fusible link made of metal that melts at a fairly low temperature when the air surrounding it is heated by a fire. Sprinkler systems have proved to be a highly reliable and effective means of fire suppression. Smoke can be as dangerous as fire to building occupants, and protective measures include the automatic shutdown of mechanical ventilating systems and the division of the building into smoke proof compartments to prevent the spread of smoke.

The evacuation of occupants in emergencies is accomplished by a system of protected exits leading to the exterior; all building areas must be within a specified travel distance of such an exit, varying from 30 to 90 metres (100 to 300 feet). For one-story buildings the exit usually consists simply of exterior doors, but for multistory buildings the exits are enclosed stairways that also lead to the exterior. The stairways have fire-rated enclosures and are often pressurized to exclude smoke; their width is determined by the maximum predicted number of occupants per floor. Travel paths to the exit must be clearly marked by illuminated directional exit signs, and battery-powered emergency lighting is required in the travel path and in the exit itself, in case of power failure. Some buildings of this type, such as hospitals, have large diesel- or natural gas-powered emergency electric generating systems that provide power and lighting for critical areas (such as operating rooms).

Another of the life-safety elements in these buildings is the fire-resistance requirements for building materials. These include the application of cementitious fireproofing or insulation to structural steel frames, the fire-resistive construction of the enclosures around exits, the flame-spread ratings of finish materials such as carpeting and wall coverings, and the use of such inherently fire-resistant materials as reinforced concrete and heavy timber. The fire-resistive ratings of various construction materials and assemblies are established by laboratory fire tests.

Vertical Transportation

Vertical transportation systems in these low buildings include stairways, sometimes only those provided as life-safety exits but more often open, well-lighted ones as well. Where large numbers of people need to be moved vertically a short distance, escalators, or moving stairways, powered by electric motors are often provided. For moving smaller volumes of people and freight, hydraulic elevators are used; the cabs of these elevators are moved by a telescoping tubular piston underneath, which is raised and lowered by pumping oil in and out of it with an electric pump. Hydraulic elevators move slowly, but they are the least expensive type and are well suited for low buildings.

Plumbing

Plumbing systems for water supply and wastewater removal are very similar to those used in residential buildings, but the higher population densities of commercial, institutional, and industrial buildings require larger toilet rooms for public multiperson use. These often include pressure-valve water closets placed in partitioned cubicles and urinals in men's toilet rooms. Some fixtures in each toilet room must be carefully arranged for easy access by handicapped persons.

The internal drainage of large flat roofs introduces another piping system, similar to that for sanitary wastewater, to carry away storm water to separate underground storm sewers. Heavy rainstorms can introduce huge influxes of water into storm sewers, and sometimes this surge effect is tempered by the use of storm water retention ponds on the building site; runoff from the roof and paved areas is temporarily stored in these ponds while it flows into the sewer at a slower rate. Hospitals, laboratories, and factories have many other types of plumbing systems for various gases and liquids; these require special materials and construction. The sites of commercial, institutional, or industrial buildings may have underground networks of irrigation piping that terminate in flush sprinkler heads to water grass and plantings.

Environmental Control

The atmosphere systems of industrial buildings are usually simple, involving only winter heating and possibly humidity control if the manufacturing process is sensitive to it. A commonly used element is the unit heater, in which an electric fan blows air through a coil heated by hot water, steam, electric resistance, or gas combustion and provides a directed supply of warm air where needed. Another system involves radiant heating using electric resistance coils backed by reflectors or continuous reflector-backed metal pipes that radiate heat from gas burned inside them. Ventilation in industrial buildings is sometimes done with operable windows but more often with unit ventilators, which penetrate walls or roofs and use electric fans to exhaust interior air that is replaced by air flowing in through operable louvres.

Commercial and institutional low-rise buildings generally have fully controlled atmospheres with heating, cooling, and humidification. An economical method of providing this controlled atmosphere is with rooftop single or multizone package units. Each unit contains an electric fan to move conditioned air; heating elements, which can be gas or oil-fired or electric resistance coils; cooling coils, which use the compressive cooling cycle with compressor, cooling coils, and condensor coils to liberate heat; as well as a fresh-air intake and air exhaust. All of these elements are prefabricated in a rectangular enclosed unit that is simply set on the roof over an opening through which it is connected to the supply and exhaust ducts. The airflow over the heating and cooling elements can be partitioned to provide different conditioned airstreams to serve different zones of the building. The conditioned air is fed at a constant volume into treelike systems of insulated sheet-metal ductwork for transmission to the zones served. The conditioned air enters the occupied space through diffusers placed in the ceiling system and connected to the ducts by flexible spiral reinforced fabric tubes. Thermostats within the space sense temperatures and send signals by electricity or compressed airflow to the unit to adjust heating and cooling as required; relative humidity is held to a range of 20 to 40 percent. The return of air from the occupied space to the unit for reconditioning is sometimes done through a reverse tree of ductwork leading back to the unit, but more often in commercial buildings this is accomplished by placing the entire sandwich space between the ceiling and the structural deck above under negative pressure to make what is called a return-air plenum. The negative pressure is created by an opening into the plenum from the return side of the rooftop unit, and perforated openings or grills in the ceiling plane admit the return air from the occupied space. Return air can also be made to enter the plenum by passing over the lamps of fluorescent light fixtures; this permits the direct recovery of heat generated by the lamps, which can be recycled to the occupied space in winter.

The rooftop unit is best used in one-story buildings or smaller multistory ones. For larger multistory buildings, centralized atmosphere systems are used. These are built up of separate components, most of which are housed in mechanical equipment rooms or in penthouses at roof level.

The components include fans for moving air, humidification devices, air-filtering devices, and refrigeration machines. Where large refrigeration machines are used, the condensor coils that liberate heat are no longer placed outside the building as in residential units or rooftop units but are located in a water jacket near the compressor. This water is circulated through a piping system to carry away the heat to a cooling tower outside the building where the water is sprayed into the atmosphere and partially evaporated to liberate heat, then recovered and returned at a lower temperature to the condensor coil jacket. Mechanical equipment rooms for atmosphere systems require a minimum of 5 percent of the floor space in a commercial building and can range up to 20 percent in hospitals and 40 percent in laboratory buildings; if the building is large, there can be more than one fan room with centralized refrigeration machines and cooling towers. The distribution of conditioned air in buildings with centralized atmosphere systems is usually done through an insulated ductwork tree, using the variable air volume (VAV) method. This method supplies conditioned air in variable amounts as required maintaining the desired temperature in occupied spaces; it results in considerable energy economies over constant volume air supply methods. Separate exhaust systems are used for areas generating heat and odours, such as kitchens, laboratories, and toilet rooms.

Electrical Systems

Electrical systems in these buildings begin at a step-down transformer provided by the utility company and located within or very close to the building. The transformer reduces the standard line potential to two dual voltage systems, which then pass through master switches and electric meters to record the subscriber's usage. Each of the voltages provided serves a separate category of use; different levels are required for incandescent lights and small appliances, large appliances, ceiling-mounted non-incandescent lighting, and heavy machinery. Each voltage pair has a separate distribution system of wiring leading from the meters and master switches to circuit breaker panels, where it is further broken down into circuits similar to residential uses. Because high-voltage wiring is considered hazardous, the switches controlling overhead lighting use lower voltages, and each heavy machine has its own fused switch. From the circuit breaker panel, low-voltage power conduit and wiring is typically distributed through partitions and ceiling sandwich spaces, but, in large open areas of commercial buildings, there may be wireways embedded in the floor slab. These wireways can be either rectangular metal tube inserted into the concrete slab before pouring or closed cells of formed steel deck; the wireways are tapped where desired to provide convenience outlets at floor level.

Lighting in these buildings is predominantly fluorescent. Lamps range in size and wattage, and the available colours can range from warm white to cool white. Incandescent tungsten-filament lamps are used mostly for accent lighting, since their light-output efficiency is low. Mercury-vapour and metal halide-vapour lamps have the same efficiency as fluorescent lamps, but certain types may have longer operating lives. High-pressure sodium-vapour lamps have even higher efficiencies and are used in industrial applications; their marked orange colour and high intensity has limited their commercial and institutional use, however. Each of these types of lamp is used in a variety of fixtures to produce different lighting conditions. Incandescent lamps can be placed in translucent glass globes for diffuse effects, or in recessed ceiling-mounted fixtures with various types of reflectors to evenly light walls or floors. Fluorescent lamps are typically installed in recessed rectangular fixtures with clear prismatic lenses, but there are many other fixture types, including indirect cove lights and luminous ceilings with lamps placed above suspended plastic or metal eggcrate diffuser grids. Mercury-vapour and high-pressure sodium-vapour lamps are placed in simple reflectors in high-ceilinged industrial spaces, in pole-mounted light fixtures for outdoor applications on parking lots and roadways, and in indirect up-lighting fixtures for commercial applications.

Mathematical models can accurately predict the performance of lighting in most applications. The zonal cavity method, which takes into account the lamps, fixtures, shape of room, and colours of room surfaces, is one example. The usual measure of light intensity is in footcandles on a horizontal surface, such as the floor of a room or a desk. The intensity ranges from 15 footcandles for a minimum ambient light level to 70 footcandles for an office or classroom and 100–200 footcandles for very precise visual tasks such as drafting; direct sunlight at noon, by comparison, is about 1,000 footcandles. In most of these buildings, the required lighting level is achieved with fixtures mounted at ceiling level; having all lighting at ceiling level allows flexibility in using building spaces. But the intensity of light varies inversely with the square of the distance from the source; thus, if a light fixture gives an intensity of 40 footcandles at a distance of one metre, it will produce an intensity of 10 footcandles at two metres. Therefore, considerable energy savings can be realized by having a minimal ambient light level (say 15 footcandles) produced by ceiling-mounted fixtures and providing task lighting close to work surfaces where higher intensities are needed. Daylighting from windows and skylights is also utilized in these buildings, and mathematical models have been developed that accurately predict its performance.

Communications systems are of growing significance and complexity in commercial, institutional, and industrial buildings. Thus communications wires for telephones, public-address systems, and computer data are free to take many paths through the building, including vertical risers, ceiling sandwich spaces, and wireways in floor slabs similar to those of electrical power wires. Where the density of wires rises to very high levels—for example, in computer rooms or where many small computer terminals are installed—raised floor systems are used. Removable floor panels are mounted on tubular metal frameworks resting on the structural floor slab, creating a plenum space to carry the necessary wiring.

A number of building systems are controlled by computers or microprocessors. In certain atmosphere systems both the interior sensors (such as thermostats) and the exterior weather sensors feed data to a computer that adjusts the system for minimal energy expenditure. Other examples include security, fire, and emergency alarm systems.

High-Rise Buildings

The high-rise building is generally defined as one that is taller than the maximum height which people are willing to walk up; it thus requires mechanical vertical transportation. This includes a rather limited range of building uses, primarily residential apartments, hotels, and office buildings, though occasionally including retail and educational facilities. A type that has appeared recently is the mixed-use building, which contains varying amounts of residential, office, hotel, or commercial space. High-rise buildings are among the largest buildings built, and their unit costs are relatively high; their commercial and office functions require a high degree of flexibility.

Foundations

The foundations of high-rise buildings support very heavy loads, but the systems developed for low-rise buildings are used, though enlarged in scale. These include concrete caisson columns bearing on rock or building on exposed rock itself. Bearing piles and floating foundations are also used.

Structural Systems

Wind Loads

The structural systems of tall buildings must carry vertical gravity loads, but lateral loads, such as those due to wind and earthquakes, are also a major consideration. Maximum 100-year-interval wind forces differ considerably with location; in the interiors of continents they are typically about 100 kilograms per square metre (20 pounds per square foot) at ground level. In coastal areas, where cyclonic storms such as hurricanes and typhoons occur, maximum forces are higher, ranging upward from about 250 kilograms per square metre (50 pounds per square foot). Wind forces also increase with building height to a constant or gradient value as the effect of ground friction diminishes. The maximum design wind forces in tall buildings are about 840 kilograms per square metre (170 pounds per square foot) in typhoon areas.

The effect of wind forces on tall buildings is twofold. A tall building may be thought of as a cantilever beam with its fixed end at the ground; the pressure of the wind on the building causes it to bend with the maximum deflection at the top. In addition, the flow of wind past the building produces vortices near the corners on the leeward side; these vortices are unstable and every minute or so they break away downwind, alternating from one side to another. The change of pressure as a vortex breaks away imparts a sway, or periodic motion, to the building perpendicular to the direction of the wind. Thus, under wind forces there are several performance criteria that a high-rise structure must meet. The first is stability—the building must not topple over; second, the deflection, or sidesway at the top, must not exceed a maximum value (usually taken as $^1/_{500}$ of the height) to avoid damage to brittle building elements such as partitions; and, third, the swaying motion due to vortex shedding must not be readily perceptible to the building occupants in the form of acceleration, usually stated as a fraction of gravity, or g. The threshold of perception of lateral motion varies considerably with individuals; a small proportion of the population can sense 0.003 g or 0.004 g. The recommendation for motion perception is to limit acceleration to 0.010 g for wind forces that would recur in 10-year intervals. The fourth criterion involves the natural period of the building structure. This is the vibration period at which the swaying cantilever motions of the building naturally reinforce and enhance each other and could become large enough to damage the building or even cause it to collapse. The natural period of the building should be less than one minute, which is the period of vibration due to the shedding of wind vortexes.

Earthquake Loads

Earthquake or seismic forces, unlike wind forces, are generally confined to relatively small areas, primarily along the edges of the slowly moving continental plates that form the Earth's crust. When abrupt movements of the edges of these plates occur, the energy released propagates waves through the crust; this wave motion of the Earth is imparted to buildings resting on it. Timber frame buildings are light and flexible and are usually little damaged by earthquakes; masonry buildings are heavy and brittle and are susceptible to severe damage. Continuous frames of steel or reinforced concrete fall between these extremes in their seismic response, and they can be designed to survive with relatively little damage.

In two major earthquakes involving large numbers of high-rise buildings, in Los Angeles in 1971 and Mexico City in 1985, lateral accelerations due to ground motions in a number of tall buildings were measured with accelerometers and were found to be of the order of 0.100 to 0.200 g. In Los Angeles, where the period of the seismic waves was less than one second, most steel-frame high rises performed well with relatively little damage; continuous concrete frames also generally performed well, but there was considerable cracking of concrete, which was later repaired by the injection of epoxy adhesive. In Mexico City, however, the period of the seismic waves was quite long, on the order of a few seconds. This approached the natural frequency of many tall structures, inducing large sidesway motions that led to their collapse. Based on this experience, determination of the seismic performance criteria of buildings involves the lateral resistance of forces of 0.100 to 0.200 g and consideration of the natural period of the building in relation to the period of seismic waves that can be expected in the locality. Another important factor is the ductility of the structure, the flexibility that allows it to move and absorb the energy of the seismic forces without serious damage. The design of buildings for seismic forces remains a complex subject, however, and there are many other important criteria involved.

Classification of Structural Systems

The types of structures used for high-rise buildings must meet the lateral load performance criteria outlined above, and they must be reasonably efficient in the use of material and of reasonable cost. The most efficient high-rise structure would meet the lateral load criteria using no more material than would be required for carrying the building gravity load alone; in other words, it would have no premium for height. This economic criterion of "no premium for height" has led to a classification of high-rise structures, each of which has only a small premium for a particular range of height.

High-rise structures begin at the lowest range with the rigid frame in both steel and concrete. Some or all of the joints between the beams and columns are rigidly joined together by welding the steel or pouring the concrete in situ, and lateral resistance is provided by the rigid joints; this system can rise about 90 metres (300 feet) with little premium. The next type is the rigid frame with a vertical shear truss in steel or a shear wall in concrete to provide greater lateral rigidity; it has a range of 38 to 150 metres (125 to 500 feet). The framed tube structure in both steel and concrete brings more gravity load and more structural material to closely spaced columns at the building's perimeter, again increasing lateral rigidity; this type is reasonably efficient from 38 to 300 metres (125 to 1,000 feet) in height. The trussed tube with interior columns, which can also be executed in both steel and concrete, introduces diagonal bracing on all sides of the building's perimeter. The bracing also carries gravity loads and further raises the lateral rigidity, making this a low-premium structure for the region of 240 to 360 metres (800 to 1,200 feet). The bundled tube, which consists of a number of framed tubes joined together for even greater lateral rigidity, begins to be practical at about 75 metres (250 feet). It was the form of the steel structure used for the Sears (now Willis) Tower in Chicago. Beyond this height there is another system that appears to have a low premium: the superframe. In this structure much of the building's gravity load, and therefore its material, is brought to a diagonally braced superframe tube at the perimeter by interior transfer trusses of various configurations. No true superframes have yet been built.

Enclosure Systems

The enclosure systems for high-rise buildings are usually curtain walls similar to those of low-rise buildings. The higher wind pressures and the effects of vortex shedding, however, require thicker glazing and more attention to sealants. The larger extent of enclosed surfaces also requires consideration of thermal movements, and wind- and seismic-induced movements must be accommodated. Window washing in large buildings with fixed glass is another concern, and curtain walls must provide fixed vertical tracks or other attachments for window-washing platforms. Interior finishes in high-rise buildings closely resemble those used in low-rise structures.

Life-Safety Systems

Life-safety systems are similar to those in low-rise buildings, with stairways serving as vertical emergency exits; in case of fire all elevators are automatically shut down to prevent the possibility of people becoming trapped in them. Emergency generator systems are provided to permit the operation of one elevator at a time to rescue people trapped in them by a power failure. Generators also serve other vital building functions such as emergency lighting and fire pumps. Fire-suppression systems often include sprinklers, but, if none are required by building codes, a separate piping system is provided with electric pumps to maintain pressure and to bring water to fire-hose cabinets throughout the building. There are also exterior connections at street level for portable fire-truck pumps. The fire hoses are so placed that every room is accessible; the hoses are intended primarily for professional fire fighters but may also be used by the building occupants.

Vertical Transportation

Vertical transportation systems are of vital importance in high-rise buildings. Escalators are used on lower floors for moving high volumes of people over short distances. A few retail or educational buildings have escalators for up to 10 stories. The principal means of vertical transport in tall buildings is the roped elevator. It moves by a direct current electric motor, which raises and lowers the cab in a shaft with wire ropes running over a series of sheaves at the motor and the cab itself; the ropes terminate in a sliding counterweight that moves up and down the same shaft as the cab, reducing the energy required to move the elevator. Each elevator cab is also engaged by a set of vertical guide tracks and has a flexible electric cable connected to it to power lighting and doors and to transmit control signals. Passenger elevators range in capacity from 910 to 2,275 kilograms (2,000 to 5,000 pounds) and run at speeds from 90 to 510 metres per minute; freight elevators hold up to 4,500 kilograms (10,000 pounds). The speed of elevators is apparently limited to the current value of 510 metres per minute by the acceleration passengers can accept and the rate of change of air pressure with height, which at this speed begins to cause eardrum discomfort.

Elevator movements are often controlled by a computer that responds to signals from call buttons on each floor and from floor-request buttons in each cab. The number of elevators in a building is determined by the peak number of people to be moved in a five-minute period, usually in the early morning; for example, in an office building this is often set at 13 percent of occupancy. The average waiting time for an elevator between pressing the call button and arrival must be less than 30 seconds in an office building and less than 60 seconds in an apartment building. The elevators are usually arranged in groups or banks ranging from one to 10 elevators serving a zone of floors, with no more than five elevators in a row to permit quick access by passengers. In a few very tall buildings the sky lobby system is used to save elevator-shaft space. The building is divided vertically into subbuildings, each with its own sky lobby floor. From the ground floor large express elevators carry passengers to the sky lobby floors, where they transfer to local elevator banks that take them to the individual floors within the subbuildings.

Plumbing

Plumbing systems in tall buildings are similar to those of low-rise buildings, but the domestic water-supply systems require electric pumps and tanks to maintain pressure. If the building is very tall, it may require the system to be divided into zones, each with its own pump and tank.

Environmental Control

The atmosphere systems in high-rise office buildings are similar to those of low-rise, with conditioned air distributed by a ductwork tree using the VAV system and return air removed through ceiling plenums. The placement of air-handling equipment can be done in two ways. One uses centralized fans placed about every 20 floors, with air moved vertically through trunk ducts to and from each floor; the other uses floor-by-floor fan rooms to provide air separately for each floor. There is usually a central refrigeration plant for the entire building connected with cooling towers on the roof to liberate heat. The central refrigeration machines produce chilled water, which is circulated by electric pumps in a piping system to the air-handling fans in order to cool incoming air as required. Incoming air is heated in winter either by piping coils through which hot water is circulated by pumps and piping from a central boiler, or by electric resistance coils in the air-handling units. In residential high-rise buildings cooling is typically provided by window air-conditioning units, and heating by hot-water or electric resistance radiant systems. There is limited use of centralized cooling, in which chilled water from a central refrigeration plant is circulated to fan-coil units near the building perimeter; a small electric fan within the unit circulates the air of the room over the chilled water coil to absorb heat.

Electrical Systems

Electrical systems for high-rise buildings are also very similar to low-rise types. The major difference is that, if the building is exceptionally tall, the utility company may bring its high-voltage lines inside the building to a number of step-down transformers located in mechanical equipment spaces. From each step-down transformer the distribution of electricity is similar to that of a smaller building.

Long-Span Buildings

Long-span buildings create unobstructed, column-free spaces greater than 30 metres (100 feet) for a variety of functions. These include activities where visibility is important for large audiences (auditoriums and covered stadiums), where flexibility is important (exhibition halls and certain types of manufacturing facility), and where large movable objects are housed (aircraft hangars).

In the late 20th century, durable upper limits of span have been established for these types: the largest covered stadium has a span of 204 metres (670 feet), the largest exhibition hall has a span of 216 metres (710 feet), and the largest commercial fixed-wing aircraft has a wingspread of 66.7 metres (222 feet) and a length of 69.4 metres (228 feet), requiring a 75–80-metre- (250–266-foot) span hangar. In these buildings the structural system needed to achieve these spans is a major concern.

Structural Systems

Structural Types

Structural systems for long-span buildings can be classified into two groups: those subject to bending, which have both tensile and compressive forces, and funicular structures, which experience either pure tension or pure compression. Since bridges are a common type of long-span structure, there has been interplay of development between bridges and long-span buildings. Bending structures include the girder, the two-way grid, the truss, the two-way truss, and the space truss. They have varying optimum depth-to-span ratios ranging from 1: 5 to 1 : 15 for the one-way truss to 1 : 35 to 1 : 40 for the space truss. The funicular structures include the parabolic arch, tunnel vault, and dome, which act in pure compression and which have a rise-to-span ratio of 1: 10 to 1: 2, and the cable-stayed roof, the bicycle wheel, and warped tension surfaces, which act in pure tension. Within these general forms of long-span structure, the materials used and labour required for assembly are an important constraint along with other economic factors.

Timber Structures

Glue-laminated timber can be used as a long-span material. It can be prefabricated using metal connectors into trusses that span up to 45 metres (150 feet). Its most economical forms, however, are the pure compression shapes of the multiple-arch vault, with spans up to 93 metres (305 feet), and ribbed domes, with spans up to 107 metres (350 feet). These are often used as industrial storage buildings for materials such as alumina, salt, and potash that would corrode steel or concrete. Such timber structures are usually found only near forested areas; transportation of timber to other areas increases its cost.

Steel Structures

Steel is the major material for long-span structures. Bending structures originally developed for bridges, such as plate girders and trusses, are used in long-span buildings. Plate girders are welded from steel plates to make I beams that are deeper than the standard rolled shapes and that can span up to 60 metres (200 feet); however, they are not very efficient in their use of material. Trusses are hollowed-out beams in which the stresses are channeled into slender linear members made of rolled shapes that are joined by welding or bolting into stable triangular configurations. The members of trusses act either in pure compression or pure tension: in the top and bottom horizontal members the forces are greatest at the centre of the span, and in the verticals and diagonals they are greatest at the supports. Trusses are highly efficient in bending and have been made up to 190 metres (623 feet) in span. Two-way grids can be made of either plate girders or trusses to span square spaces up to 91 metres (300 feet) in size; these two-way structures are more efficient but more expensive to build.

The highly efficient funicular forms are used for the longest spans. Vaults made of rows of parabolic arches, usually in truss form for greater rigidity, have been used for spans of up to 98.5 metres (323 feet). Steel truss domes, particularly the Schwedler triangulated dome, have been the choice for several large covered stadiums, with the greatest span being 204.2 metres (669 feet). Cable-stayed roof construction is another structural system derived from bridge building. A flat roof structure in bending is supported from above by steel cables radiating downward from masts that rise above roof level; spans of up to 72 metres (236 feet) have been built. Another funicular form is the bicycle-wheel roof, where two layers of radiating tension cables separated by small compression struts connect a small inner tension ring to the outer compression ring, which is in turn supported by columns.

Tension-cable networks use a mesh of cables stretched from masts or continuous ribs to form a taut surface of negative curvature, such as a saddle or trumpet shape; the network of cables can be replaced by synthetic fabrics to form the tension surface. Another fabric structure using tension cables is the air-supported membrane. A network of cables is attached by continuous seams to the fabric, and the assembly of cables and fabric is supported by a compression ring at the edge. The air pressure within the building is increased slightly to resist exterior wind pressure. The increase can be as slight as 1.5 percent of atmospheric pressure, and it is possible to maintain this even in large buildings with relatively small compressors. The cables stiffen the fabric against flutter under uneven wind pressure and support it in case of accidental deflation.

Concrete Structures

Reinforced concrete, because of its inherent strength in compression, is primarily used for long spans in funicular compression forms, including vaults, shells, and domes. Thin parabolic shell vaults stiffened with ribs have been built with spans up to about 90 metres (300 feet). More complex forms of concrete shells have been made, including hyperbolic paraboloids, or saddle shapes, and intersecting parabolic vaults. An example of the latter is the CNIT Exhibition Hall in Paris, which consists of six intersecting double-shell parabolic vaults built to span a triangular space 216 metres (708 feet) on a side with supports only at the apexes of the triangle. Reinforced concrete domes, which are usually also of parabolic section, are built either in ribbed form or as thin shells. The maximum span of these domes is about 200 metres (660 feet).

Another funicular form used in concrete, though it is really a composite structure, is the inverted dome, or dish. As in the steel bicycle wheel, a concrete compression ring resting on columns at the perimeter of the structure supports radial steel cables that run inward and downward to a small steel tension ring at the centre, forming the dish shape. The cable network is stiffened against wind forces by encasing it in a poured concrete dish; structures of this type have been built with spans of up to 126 metres (420 feet).

Factors in the Built Environment

Acoustics

Long-span auditoriums involve considerations in acoustics: audiences wish to hear speakers clearly and to hear music with appropriate tonality. Unfortunately, acoustic requirements for speech quality often conflict with those for music, and it is difficult to design an auditorium that is satisfactory for both. The best single measure of acoustic performance for auditoriums is the reverberation time, which is directly proportional to the volume of the hall and inversely proportional to the amount of sound absorbency within it, including wall and ceiling surfaces and the audience itself. Measured in the sound range of 500–1,000 hertz, rooms with short reverberation times of one to 1.5 seconds are good for the intelligibility of speech, while longer reverberation times of 1.5 to 2.5 seconds add richness of tone to musical performances. Thus, adding sound-absorbent material to a hall improves it for speech but detracts from its musical qualities. People are excellent sound absorbers, and thus the audience has a distinct impact on auditorium acoustics; to keep this effect constant with varying audience size, auditorium seats are usually upholstered to serve as surrogate spectators of the same sound absorbency. Curved surfaces, which tend to focus sound, are either avoided in auditoriums or covered with sound-absorbent material. Electronic sound-amplification systems can be used to assist speakers in large halls but generally are not satisfactory for music. Other long-span buildings, such as covered stadiums and exhibition halls, receive only minor acoustical treatment.

Environmental Control Systems

Atmosphere systems in long-span buildings must handle the considerable heat and odour generation from population densities of less than one square metre (11 square feet) per person. Air must be moved fairly rapidly through the population zone to maintain an acceptable air-change rate.

Construction Site Safety

Construction safety involves any safety procedure that is related to the construction industry or construction sites. Construction safety aims to ensure that a construction site or the industry as a whole is not the cause of immediate danger to the public around a construction site, or the workers at a construction site, as well as making sure that the finished product of construction meets required safety standards.

Construction safety is a major cause for concern in the working world, as it is one of the most dangerous. It is, according to the Bureau of labour statistics (BLS), the fourth most dangerous profession with the second most fatal injuries. It is possibly the second most dangerous land-based profession after the fishing industry. The Occupational Safety and Health Administration (OSHA) provides health and safety regulations and standards specific to the construction industry.

Construction Site Hazards

Construction hazards are heavily dependent on the type of construction work that is being carried out. For example, working on scaffolding presents entirely different hazards to working with asbestos.

The top ten risks and hazards from working on construction sites are:

- Working at height,
- Moving objects,

- Slips, trips, and falls,
- Noise,
- Hand arm vibration syndrome,
- Material and manual handling,
- Collapsing trenches,
- Asbestos,
- Electricity,
- Airborne fibres and materials.

Working at Height

The Health and Safety Executive (HSE) found that, in 2015/16, just over a quarter of worker fatalities were from a fall from height. This put working at height as the most common cause of fatal injuries to workers.

Suitable training is required for all employees who work at height. Employees should be trained in working on different pieces of equipment and surfaces, such as how to work safely on scaffolding, ladders, and roofs.

The law requires that all employers must assess the risks from working at height. A plan should then be devised to ensure all work is carried out safely. The employee must have received the correct training in working at height, and must be aware of the safety procedures they should follow when doing so.

Working at height must be properly planned and supervised, and certain approaches and precautions should be adopted. These are:

- Avoid working at height where possible. For example, if something can be assembled on ground level, do it there.
- Use equipment with an extra level of safety to reduce the risk of a fatal fall. For example, a scaffold with a double guard-rail.
- Minimise the consequences of a fall, for example by providing a safety net.

Moving Objects

A construction site is an ever-changing environment, and construction hazards continue to increase as construction is underway. There are many moving objects commonly encountered on construction sites. These include overhead lifting equipment, supply vehicles, and diggers, all of which move around a usually uneven terrain.

Reducing risks should always be a priority. Workers should always:

- Avoid working close to the moving object.
- Be vigilant of their surroundings, especially if the object does not have lights or beepers.
- Wear Personal Protective Equipment (PPE), such as a high visibility jacket, to ensure they are seen.

Slips, Trips and Falls

Slips, trips, and falls can happen in almost any environment. As construction sites often have uneven terrain, buildings at various stages of completion, and unused materials on site, it is unsurprising that slips, trips, and falls are a common hazard.

HSE reports that several thousand construction workers are injured every year following a slip or trip, and that most of these could be avoided by effectively managing working areas and access routes, such as stairwells and footpaths.

Those in control of construction sites must effectively manage the site so that workers can move around it safely. Risks should always be reported and sorted to reduce the chances of injury.

Some causes of slips and trips and how to prevent them include:

- Uneven surfaces – The risk of these can be reduced by providing walkways that are clearly designated as walkways, having good conditions underfoot, and being well lit.
- Obstacles – Instances of slipping and tripping over obstacles can be dramatically reduced by everyone keeping their work and storage areas tidy and designating specific areas for waste collection.
- Trailing cables – Cordless tools should be used where possible. If this is not possible, cables should be run at high levels.
- Wet or slippery surfaces – If a surface is slippery with mud it should be treated with stone, and if it is slippery with ice it should be treated with grit. Any areas that are slippery should be signposted, and footwear with a good grip should be worn.

Noise

Construction is noisy and, as a result, noise is a common construction hazard. Loud, repetitive, and excessive noise causes long term hearing problems, such as deafness. Noise can also be a dangerous distraction and may distract the worker from the task at hand, which can cause accidents.

It is the employer's responsibility to carry out a comprehensive noise risk assessment, and provide appropriate PPE where necessary.

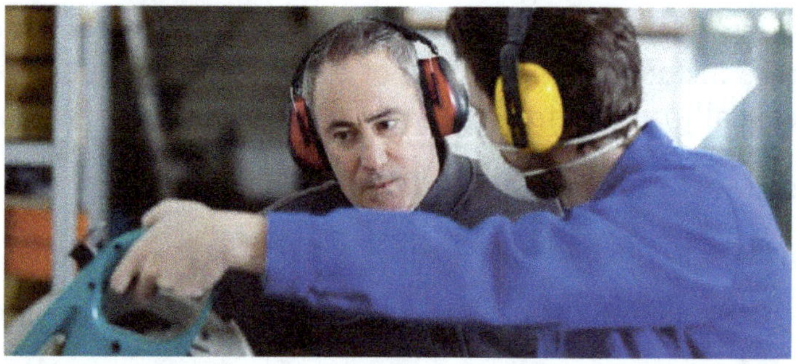

Hand Arm Vibration Syndrome

Hand Arm Vibration Syndrome (HAVS) is a painful and debilitating disease of the blood vessels, nerves, and joints. It is usually caused by the prolonged use of hand-held power tools, including vibratory power tools and ground working equipment.

HAVS is preventable, however once the damage is done, it is permanent. HSE reports that nearly 2 million people are at risk of developing HAVS. Damage from the disease can include the inability to do fine work, and cold temperatures can trigger painful attacks in the fingers.

Construction workers should be given appropriate protection when using vibrating tools, and equipment should be well maintained.

Material and Manual Handling

Materials and equipment are constantly being lifted and moved around construction sites, whether this be manually or by equipment. Either way, handling carries a degree of risk.

Where duties involve manual handling, adequate training must be provided. If an employee is required to use lifting equipment, they must be trained in how to use this and a test should be taken to check their ability to use the equipment safely.

Collapsing Trenches

A common occurrence on construction sites is the collapsing of trenches with workers inside. Further, a building that is being demolished or under construction can suddenly and unexpectedly collapse, which can seriously injure, or even kill, those inside.

Precautions for collapse need to be taken before work starts. If the project requires a trench, site managers should:

- Consider the kind of support that is best suited for the trench.
- Ensure the trench is fully secure.
- Regularly inspect the trench both before and during the work shift.

Asbestos

Asbestos refers to a set of six naturally occurring fibrous minerals. When materials that contain asbestos are disturbed or damaged, these fibres are released into the air. Inhaling these fibres can cause fatal and serious diseases such as lung cancer, asbestosis, and pleural thickening.

Asbestos kills around 5,000 workers per year, and an average of 20 tradespeople die every week as a result of past exposure. An estimated 500,000 public buildings in the UK are thought to contain asbestos.

If there is asbestos on the construction site, workers must be informed where it is. They must be trained in what to do should they come across suspicious materials that may contain asbestos.

Electricity

It is harmful to be exposed to electrical live parts. Harm can occur either by touching live parts directly or indirectly by a conducting object or material.

HSE reports that 1,000 electrical accidents at work are reported every year. Most of these accidents arise from contact with overhead or underground power cables and electrical equipment/ machinery. Electric shocks are a common cause for falls from ladders, scaffolds, and other work platforms. There are also a growing number of electrocutions involving workers who are not qualified electricians, but who are carrying out electrical work on construction sites.

Airborne Fibres and Materials

Unsurprisingly, a lot of dust is produced on construction sites. The dust on construction sites is often an invisible, fine, and toxic mixture of hazardous materials and fibres. This can damage the lungs and lead to diseases such as chronic obstructive pulmonary disease, asthma, and silicosis.

It is the duty of all employers to ensure protective equipment is used. Simply providing it is not enough.

Practices to Improve Construction Site Safety

Here are eight ways construction businesses can reduce workplace accidents and promote construction site safety.

Awareness

Before any worker — no matter his or her role or experience level — can set foot on a construction site, he or she must be fully aware of the possible hazards. Ignorant workers are perhaps the biggest dangers in any industry, as their unknowing mistakes put everyone else at risk. Understanding of perils at hand and sustaining a perpetual state of alertness is perhaps the number-one best way to prevent accidents. To become aware of such risks and how to avoid them, every single person that steps foot onto a construction site should be aware of the risks associated with the job and how to prevent them with their knowledge of construction site safety.

It is the construction managers' job to make sure that every worker is aware of the dangers that come with working on a construction site and they must protect workers from these dangers. Any manager that fails to tell their crew and staff about how to avoid getting hurt and how to ensure safety is failing as a manager. Ensuring the safety of the construction workers and everyone on the site should be the number one priority of any construction manager. If the workers have no concept of construction site safety, they shouldn't be allowed on the construction site.

Training

Though most of a construction worker's skills can be gained on the job, safety is one skill set that is best learned before works enter the construction site. The Occupational Safety and Health Administration (OSHA) and other organizations publish some resources to help businesses train their new laborers on standard safety and security practices, including pamphlets, worksheets, training videos, and even on-site training opportunities. Experienced workers should be expected to refresh their knowledge of standard safety by attending regular training sessions throughout the year. These training sessions can go over simple things such as fall protection and proper use of ladders, but the goal is to make sure everyone is adequately trained. Leaving these training sessions, workers should know what safety measure to do in the case of an incident.

Although workers are expected to attend regular safety training sessions about construction safety throughout the year, being able to practice safety training skills on-site would help construction workers enforce the safety rules. Practicing construction site safety training skills on-site will force workers to practice these skills in an environment where safety is essential and will make sure they are trained. Without the proper training, construction workers can be easily injured or even killed. In such an environment where injuries and fatalities are highly likely, training is necessary and will prevent things like workplace injuries that will hurt you ethically and economically.

Communication

Accidents are more likely to occur when workers are unsure what to expect. Direct discussion of the day's goals and activities will cut down on surprises that could cause bodily harm. Construction firms would be wise to equip workers with devices, like smartphones, walkie-talkies, or headsets, which allow fast and efficient communication among team members. Without proper communication between everyone on the construction site, workers won't know what to expect. Clear and concise communication with everyone not only makes the project go by faster but also helps keep each person informed. Informing the staff and making sure everyone is doing their job is a proper way to communicate and make sure they understand construction site safety.

Documentation

To enforce construction site safety, you have to make sure you have proper documentation of everything that is going to be done on-site. There are some legal hoops most construction companies must jump through to begin building, and it is essential that all proper registrations and licenses are earned before work begins. Supervisors and contractors who will be charged with particularly difficult tasks, like blasting, certainly should provide evidence of their certification well in advance of their employment on the job site. Not only does this prevent accidents due to improper training, but it protects the construction firm from legal action and public scrutiny. Any safety hazards that make their way to the media will look bad for construction firms.

No construction worker wants to work for a construction firm that doesn't put its worker's safety first. Any news of workers getting hurt on the job due to lack of safety practices will have new prospects running away from your construction firm. Implementing measures to practice construction site safety methods will prevent such things from happening. Also, documenting all work in the field using cloud and mobile technology is making it easier than ever before to mitigate future lawsuits.

Proper Equipment

To create a culture centered on construction site safety, you need to give workers the proper equipment and adequate work area for the job at hand. Without the proper equipment, you can't have construction site safety because there will always be an opportunity to get injured using the wrong equipment. Construction workers equipped with improper gear are bound to make fatal errors. Not only should each piece of equipment on the job site be ideally suited to the task at hand, but construction firms have to make sure that all machinery and material are well maintained.

Construction companies must also consider equipment that doesn't directly contribute to the construction project. Workers should have plenty of water available on-site as well as a shady place to prevent dehydration and exposure-related illnesses. Longer construction projects may even benefit from fabric structures to store equipment and cover incomplete sites. Such simple things can be easily overlooked, and if they are, they increase the chance of on-site injury. Proper construction equipment ensures that there is at least some level of construction site safety within the construction firm.

Supervision

Ideally, construction workers would fully understand the ramifications of inadequate safety precautions and thus act in a manner to ensure site-wide well-being – but this is not a perfect world. Every site must have a strong supervisor who is willing and capable of enforcing safety standards with no exceptions. This foreman must keep tabs on all employees throughout the day and correct those who fail to commit to proper construction site safety procedures.

Innovation

The accident rate would be even higher than it is today if it wasn't for construction firms willing to devote extra resources to keep their employees safe. These additional resources not only lower the rates of workplace accidents and injuries but also help develop new ideas for keeping construction workers safe. The development of new practices that will enhance security should always be encouraged, and companies should avoid speaking against legislation aimed at improving safety protocols. Perhaps with enough innovation, all construction sites can maximize their construction site safety practices and can be 100 percent accident-free.

Transparency

The worst thing any construction firm can do for its reputation is attempting a cover-up. Hiding accidents from the press and the public not only lowers the opinion of a single endeavor – it paints the building industry as a whole in a negative light. Ultimately, people understand that accidents happen, and as long as contractors are doing their best to foster a safe environment for their workers, any accidents that do occur will only contribute to the growing need to augment modern safety techniques. Transparency, along with the other seven practices on this list, will help construction as a whole become a safer industry in which to work. A safer construction industry is an industry of fewer injuries, fewer workplace accidents, and fewer deaths. A construction industry that fully utilizes its construction site safety practices is the kind of industry we should be working towards.

Building Materials Properties

1. They must last over time, which means that they must be impervious to biological and chemical attack. This applies not only to their strength, but also to their appearance and color. Many building materials are known to discolour after years of exposure to the sun's ultraviolet (UV) rays. Therefore those materials used in construction must be UV stable. Materials that are only used indoors need not be UV stable.

   Concrete can lose its strength over time if chemicals seep into it via rainwater or atmospheric pollution. It is especially susceptible to chloride attack. Steel can rust or corrode over time. Rusting has a major impact on the load carrying ability of a member, and can quickly make a structural member useless.

2. They must be strong: Every part of a building is subject to some level of force, and each part must be capable of carrying the loads that fall on it. Many building components have other components attached to them, like brackets, railings, light fixtures, pipes, ducts, and wall shelves. They must be able to hold on to and retain these attachments. Building materials can be rated for their compressive strength (think of bricks pressing down on each other), or tensile strength (like the force you exert on the chain of a swing when you stand on it), or both.

3. They must resist wear and tear: In the building industry, wear and tear is called abrasion resistance. It is especially important for areas like floor finishes, but every part of the building is subject to some level of abrasion, and must be able to resist this. You can actually measure how abrasion resistant a building material is by performing a standardized test such as a Taber Test. In this test, a metal disc of standardized design and weight is rotated against the material to be tested, at a standard speed for a fixed length of time. The material sample is weighed before and after the test. The difference in weight gives us an indication of how much material was lost during the test. The more the material lost, the less the abrasion resistance of the material. In this way you can compare, say, the abrasion resistance of a marble floor versus a ceramic tile and then decide which works best for that particular area.

References

- Construction, technology: britannica.com, Retrieved 15 March, 2019
- Modern-building-practices, construction, technology: britannica.com, Retrieved 4 May, 2019
- Construction-safety-3970: safeopedia.com, Retrieved 14 January, 2019
- Common-construction-hazards, hub: highspeedtraining.co.uk, Retrieved 4 June, 2019
- Improve-construction-site-safety: esub.com, Retrieved 12 April, 2019
- Building-materials: understandconstruction.com, Retrieved 20 February, 2019

2

Building Materials and their Properties

The different materials which are used for the purpose of construction are known as building materials. Some of the different types of building materials are mud and clay, fabric and natural construction materials. The chapter closely examines these types of building materials as well as their properties to provide an extensive understanding of the subject.

Building material is any material used for construction purpose such as materials for house building. Wood, cement, aggregates, metals, bricks, concrete, clay are the most common type of building material used in construction. The choices of these are based on their cost effectiveness for building projects.

Many naturally occurring substances, such as clay, sand, wood and rocks, even twigs and leaves have been used to construct buildings. Apart from naturally occurring materials, many man-made products are in use, some more and some less synthetic.

The manufacture of building materials is an established industry in many countries and the use of these materials is typically segmented into specific specialty trades, such as carpentry, plumbing, roofing and insulation work. This reference deals with habitats and structures including homes.

A wide range of building materials is available for the construction of rural buildings and structures. The proper selection of materials to be used in a particular building or structure can influence the original cost, maintenance, ease of cleaning, durability and, of course, appearance.

Several factors need to be considered when choosing the materials for a construction job, including:

1. Type and function of the building or structure and the specific characteristics required of the materials used, i.e. great strength, water resistance, wear resistance, attractive appearance, etc.

2. Economic aspects of the building/structure in terms of original investment and annual cost of maintenance.
3. Availability of materials in the area.
4. Availability of the skilled labour required to install some types of material.
5. Quality and durability of different types of material.
6. Transportation costs.
7. Selection of materials with compatible properties, dimensions and means of installation.
8. Cultural acceptability or personal preference.

Wood

Wood is a commonly used construction material in many parts of the world because of its reasonable cost, ease of working, attractive appearance and adequate life if protected from moisture and insects. However, forests are a valuable natural resource that must be conserved, particularly in areas with marginal rainfall. As good a material as wood may be, there are regions where other materials should be considered first, simply on a conservation basis.

Wood for building is available from many different species with widely varying characteristics. Some species are used in the form of small poles for light construction, while other species are allowed to mature so that timber (lumber in many countries) may be sawn from the large logs. The species that produce small, inexpensive poles in rather short growing periods often grow in the fringes of agricultural land and can be used without danger to the ecology of the region.

Hardwoods versus Softwoods

Wood cut from deciduous trees (which drop their leaves sometime during the year) is considered to be hardwood, while that cut from coniferous (needle bearing) trees is considered to be softwood. However, this classification does not accurately reflect whether the wood itself is soft or hard.

Defects in Wood

Defects to watch for when selecting timber are:

Brittle heart, found near the centre of many tropical trees, makes the wood break with a brittle fracture.

Wide growth rings indicate rapid growth resulting in thin-walled fibres with consequent loss of density and strength.

Fissures include checks, splits, shakes and resin pockets. Knots are the part of a branch that has become enclosed in a growing tree. Dead knots are often loose and therefore reduce the effective area that can take tensile stress. Knots can also deflect the fibres, thereby reducing strength in tension.

Decay, which results from moisture levels between 21 percent and 25 percent in the presence of air, reduces the strength of the wood and spoils its appearance.

Insect damage caused by borers or termites.

The fungi that feed on wood can be divided into three main categories: staining fungi, moulds and decay fungi. Most fungi thrive under moist conditions. Staining fungi live mainly on the sapwood but they may penetrate deeply into the wood and spoil the timber's attractive appearance. Moulds do not penetrate below the surface and do not seem to affect the strength of the wood, but they look unsightly. Decay fungi eat the cell walls of the wood. This causes the tree to lose its strength and often reduces it to a crumbling, rotting mass. These decay fungi never attack timber that is seasoned to a moisture content of less than 20 percent and kept well ventilated and dry.

The main species of borer that attack tropical woods are the pinhole borer and the Lyctus, or powderpost beetle. The pinhole borer attacks newly felled logs and sometimes standing trees. The attack can occur within hours of felling. The beetles do not normally continue to operate in seasoned timber. The powderpost beetle attacks seasoned tropical hardwoods – particularly those that contain starch on which the larvae feed. Timber is sometimes sprayed in the yard to protect it until it is transported.

Termites are normally of two kinds: the dry wood types that are able to fly and the subterranean type. Termites usually operate under cover and it is only after the first signs of damage appear that the full extent is realized. Flying termites usually enter the end-grain of untreated timber and build up a colony from inside, finally devouring all the interior wood and leaving only a thin skin behind. Some subterranean termites, white ants, operate from a central colony and travel in search of food. Their nests or hills sometimes achieve great size and house millions of ants. While no timber is completely immune to attack from ants or other insects, there are great variations among the species. The density of the timber is no guide to its resistance to termite damage, as some of the lighter timbers are more immune than heavier varieties.

Weathering is the disintegration of wood caused by alternate shrinkage and swelling as a result of rain, rapid changes in temperature, humidity and the action of sunlight. Painting, when properly carried out, does much to prevent weathering. The paint must be of exterior quality, however, and applied according to the maker's instructions.

Poles and Timber

Wooden Poles

In farm buildings and rural structures, wood is often used in the form in which it has grown, i.e. round poles. In some areas where enough trees are grown on the farm or in local forests, wooden poles can be obtained at very low cost. These poles have many uses in small building construction, such as columns for the loadbearing structure, rafters, trusses and purlins. Sticks and thin poles are often used as wall material or as a framework in mud walls.

Where straight poles are selected for construction, it is as easy to work with round timber as with sawn timber. However, somewhat crooked poles can also be used if they are turned and twisted and put into positions in which the effects of the bends are unimportant.

Round timber can generally be considered stronger than sawn timber of the same section area because the fibres in round timber are intact. The pole is normally tapered and therefore the smallest section, the top end, must be used in the calculation of compressive and tensile strength.

A great number of species can be considered when selecting poles for construction, but only a limited number are available on the commercial market. Some species are more suitable for silviculture (growing on farms) and silvipasture (growing on pastures) than others, but must always be selected to suit local climatic and soil conditions. Generally there are several species suitable for each location that are fast and straightgrowing, and produce strong and durable timber. In addition to building poles or timber, some species will produce fodder for animals, fruit, fuelwood, etc.

Many species of eucalyptus, from which gum poles are obtained, are very fast and straight-growing hardwoods. However, they warp and split easily. Dimensions suitable for building construction are obtained by harvesting the still immature trees. Gum poles provide a strong and durable material if chemically treated.

In high-altitude areas, several species of acacia produce good building poles. Acacia melanoxylon (Australian blackwood) is very resistant to attack by termites, but grows more slowly than eucalyptus. In low- to medium-altitude areas with sandy soils and low rainfall, Casuarina produces straight and durable poles.

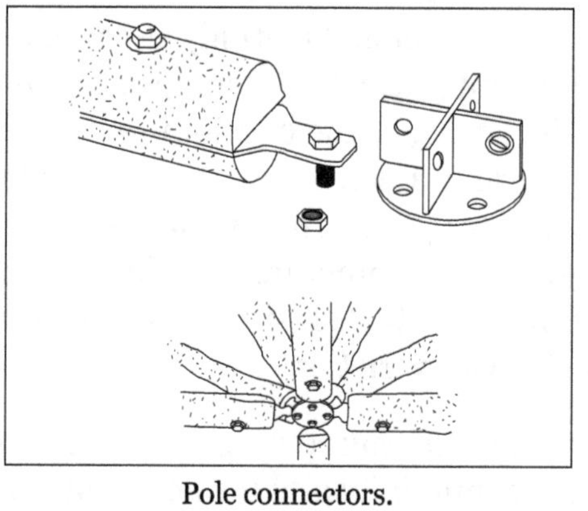

Pole connectors.

Cedar posts for fencing are obtained by splitting large logs. The posts are durable and resistant to rot and attack by termites. They are also suitable for wall posts in the construction of buildings.

In coastal areas, mangrove poles are widely used for posts in walls and trusses in roofs.

Unprocessed round wood material can be joined by being nailed or tied with string or wire. A special connector has been developed to join round wood in trusses where several members may have to be connected at each point.

Sawing Timber

The rate at which a tree grows varies with the season. The resulting growth rings of alternate high and low density form the grain in the sawn timber (lumber). The method of sawing has a considerable effect on the appearance, resistance to warping, shrinking, paintholding ability and wear resistance of the final piece.

There are several methods of sawing a log into boards and planks, giving different relationships between the growth rings and the surface, i.e. more or less parallel to the surface in plain sawn timber and at right angles in radial sawn timber.

Radially sawn boards shrink less, are less liable to cup and twist and are easier to season. Unfortunately, cutting methods that produce a high proportion of quarter-sawn timber are wasteful and therefore only used to produce material for high-class joinery work.

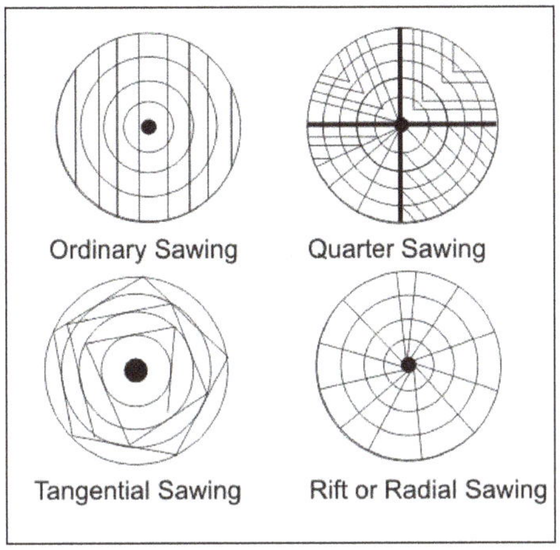

Methods of sawing timber.

Offcuts

A tree is tapered and cylindrical, whereas boards and planks are rectangular. This results in the outer pieces with tapered edges and less than full dimensions throughout the length. Such pieces, called offcuts, can sometimes be obtained at low cost and used for rough building.

Seasoning of Timber

The strength, stiffness and dimensional stability of wood are related to its moisture content. Hence, if wood is dried (seasoned) before use, not only can higher strength values be used in a design, but a more durable structure will result. In developing countries, most timber is not seasoned and it is sold in what is called its 'green' state.

Timber must be stacked, supported and sometimes restrained so as to minimize distortion during seasoning. If drying is too rapid, the outer parts, in particular the unprotected ends, shrink before the interior does, and this leads to surface checking and splitting, as well as the possible extension of ring and heart shakes. Some timber species are more difficult to season satisfactorily than others.

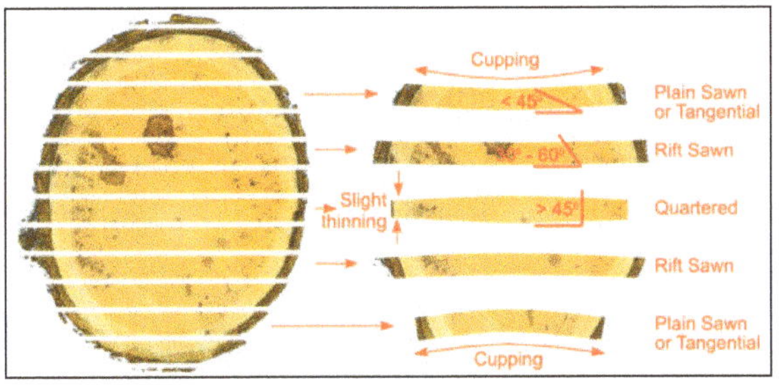

Effects of cupping of different methods of sawing.

Air Seasoning

Timber should be protected from rain and from the ground. It should therefore be stacked so that air can circulate freely around all surfaces, reducing the risk of twisting and cupping, as well as minimizing attacks by fungi and insects. In favourable conditions, thin softwoods can be air-seasoned in weeks but in unfavourable conditions some hardwoods require a year or more.

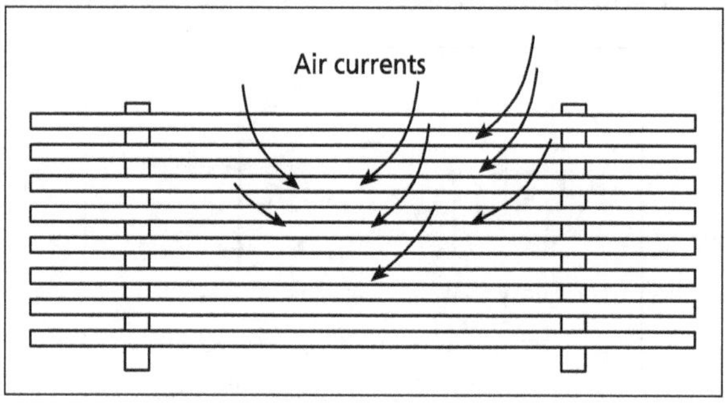

Air-drying of timber.

Artificial Seasoning

Artificial seasoning can be either moderate or rapid, depending on the temperature of the air injected into the chamber where the timber is piled, and on the rate at which the air is circulated and extracted from the chamber. This method is expensive and can only be applied on small quantities of timber. Timber can be artificially seasoned from the green condition, but often hot-air seasoning is used only at a later stage, after most of the moisture has been removed by air seasoning.

Smoke seasoning is a moderate process and involves placing the timber over a bonfire. It can take a month or two, depending on the size and type of wood being seasoned. This method is considered to be both a seasoning and a treatment method for timber. Presumably it protects the timber against pest attacks and increases durability. However, it is not very reliable and can lead to splitting of the timber because of the lack of control over the heat from the bonfire.

Care of Seasoned Timber

Timber should be protected from moisture on the building site. Close piling and covering with tarpaulins delays the absorption of atmospheric moisture, particularly in the interior of the pile.

Grades and Sizes for Timber

Grade

Grades are established by various government agencies. Even within a single country, more than one grading system may be in use. While the grade may not be important for small construction jobs, in large projects where materials are bought by specification, it is important to indicate the required grade.

Grades that provide specific information in structural design are most useful. The grade standard established by the Kenya Bureau of Standards, shown in table is a good example.

Table: Timber grades and application.

Grade	Applications
F	Furniture, high-class joinery
GJ	General joinery
S-75	Structural grade, having a value of 75% of basic stress
S-50	Structural grade, having a value of 50% of basic stress
C	A general construction grade for non-stressed construction
L	A low grade for low quality work

It is the S-75 and S-50 grades that are significant in building construction.

Sizes

Timber in eastern and southeastern Africa is available in a number of Système Internationale (SI) metric sizes, but not all are available in all localities. The dimension indicates actual size as sawn. Smoothing will reduce the timber to less than dimension size.

Timber Measurement for Trade

Even though timber is normally sold by length (running metre or foot), the price may be calculated per cubic metre when sold in large quantities. Basic lengths are between 1.8 metres and 6.3 metres, although pieces longer than about 5.1 metres are scarce and costly. Timber normally comes in running lengths, that is to say, not sorted by length.

Timber Preservation

The main structural softwood timbers of eastern and southeastern Africa are not naturally durable. If used in conditions subject to fungal, insect or termite attack, they will fail after some time. To avoid this, the timber used in permanent structures should be treated with a preservative.

Effective preservation depends on the preservative and how it is applied. An effective preservative should be poisonous to fungi and insects, permanent, able to penetrate sufficiently, cheap and readily available. It should not corrode metal fastenings, nor should the timber be rendered more flammable by its use. It is sometimes desirable to have a preservative-treated surface that can be painted.

If a structure is correctly designed and built, and the moisture content of its timber does not exceed 20 percent, then a preservative treatment is generally unnecessary for protection against fungal attack. However, where the above conditions are not present, there will be a risk of fungal decay, and proper preservation is recommended.

Wood Preservatives

Creosote is an effective general-purpose preservative that is cheap and widely used for exterior work and, to a lesser degree, indoors. It is a black to brownish oil produced by the distillation of coal tar and, while it has many of the properties required of a preservative, it increases flammability, is subject to evaporation, and creosoted wood cannot be painted. It should not be used on interiors where the characteristic smell would be objectionable. Unfortunately, creosote has been found to be a carcinogen and must be used with caution.

Coal tar is not as effective a preservative as the creosote produced from it. Tar is less poisonous, does not penetrate the timber because of its viscosity, is blacker than creosote and is unsuitable for interior wood work.

Unleachable metallic salts are based mostly on copper salts. A combination of copper/chrome/arsenate is used. The copper and arsenical salts are toxic preservatives that are rendered non-leaching (cannot be washed out) by the chrome salt, which acts as a fixing agent. The timber is impregnated using a 'vacuumpressure' process. Preservation by metallic salt is being used increasingly because the treated surfaces are odourless and can be painted or glued.

Water-soluble preservatives are not satisfactory for exterior use as they are liable to be washed out of the timber by rain.

By contrast, they are very suitable for interior work as they are comparatively odourless and colourless and the timber can be painted.

Used engine oil can often be obtained free of charge, at least in small quantities. The oil contains many residual products from combustion and some of them act as preservatives, but it is not nearly as effective as commercial preservatives. It can be thinned with diesel fuel for better penetration. The combination of 40 litres of used engine oil and 1 litre of Dieldrin is a viable alternative in rural construction.

Methods of Wood Preservation

There are two categories of timber preservation methods:

Non-Pressure Methods

These are applicable for both green and dried timber (less than 30 percent MC) and include:

1. Soaking (steeping), used for small quantities of timber.
2. Hot and cold soaking: the tank with the preservative and timber is heated to nearly boiling point for 1-2 hours and allowed to cool. During the heating period, the air in the cells expands and some is expelled. When cooling, a partial vacuum develops in the cells and liquid is absorbed.
3. Steam and cold quenching.
4. Superficial methods such as painting and spraying.

To make non-pressure methods more effective, storage in a closed environment is recommended.

Pressure Methods

The treatment in pressure processes is carried out in steel cylinders, or 'retorts'. Most units conform to size limits of 2–3 metres in diameter and up to 46 metres or more in length, and are built to withstand working pressures of up to 1 720 kPa. The wood is loaded on special tram cars and moved into the retort, which is then closed and filled with the preservative.

Pressure forces preservatives into the wood until the desired amount has been absorbed. Three processes full-cell, modified full-cell and empty-cell – are commonly used. These processes are distinguished by the sequence in which vacuum and pressure are applied to the retort. The terms 'empty' and 'full' refer to the level of preservative retained in the wood cells.

The full-cell process achieves a high level of preservative retention in the wood cells, but less penetration than the empty-cell process. On the other hand, the empty-cell process achieves relatively deep penetration with less preservative retention than the full-cell process.

Full-Cell Process

The Bethel full-cell process is generally used with water-based preservatives, especially for timber that is difficult to treat and also requires high retention. The full-cell process steps are listed below:

- The wood is sealed in the treatment cylinder and an initial vacuum is applied for approximately 30 minutes to remove as much air as possible from the wood and from the cylinder.
- The preservative either heated or at ambient temperature depending on the system, enters the cylinder without breaking the vacuum.
- Afterthecy linderisfilled, the cylinderis pressurized until no more preservative will enter the wood, or until the desired preservative retention is obtained.
- At the end of the pressure period, the pressure is released and the preservative is removed from the cylinder.
- A final vacuum may be applied to remove excess preservative that would otherwise drip from the wood.

Empty-Cell Process

The empty-cell process results in deep penetration of the preservative with a relatively low net preservative retention level. If oil preservatives are used, the empty- cell process will most probably be used, provided it will yield the desired retention level. The Rueping process and the Lowry process are the two most commonly used empty-cell processes. Both use compressed air to drive out a portion of the preservative absorbed during the pressure period.

- Rueping process

In the Rueping process, compressed air is forced into the treatment cylinder containing the charge of wood, in order to fill the wood cells with air prior to preservative injection. Pressurization times vary with wood species. For some species, only a few minutes of pressurization are required, while more resistant species may require pressure periods of 30-60 minutes. Air pressures used typically range from 172 kPa to 690 kPa, depending on the net preservative retention desired and the resistance of the wood.

After the initial pressurization period, preservative is pumped into the cylinder. As the preservative enters the treatment cylinder, the air escapes into an equalizing, or Rueping, tank at a rate that maintains the pressure within the cylinder. When the treatment cylinder is filled with preservative, the pressure is raised above the initial air pressure and maintained until the wood will take no more preservative, or until enough has been absorbed to leave the desired preservative retention level after the final vacuum. After the pressure period, the preservative is removed from the cylinder and surplus preservative is removed from the wood with a final vacuum. This final vacuum may recover 20–60 percent of the gross amount of preservative injected. The retort then is unloaded, and the treated wood stored.

- Lowry process

The Lowry process is an empty-cell process without the initial air pressure. Preservative is pumped into the treatment cylinder without either an initial air pressurization or vacuum, trapping the air that is already in the wood. After the cylinder is filled with the preservative, pressure is applied and the remainder of the process is identical to the Rueping process. The advantage of the Lowry process is that full-cell equipment can be used without the accessories required for the Rueping process, such as an air compressor, an extra tank for the preservative, or a pump to force the preservative into the cylinder against the air pressure. However, both processes are used widely and successfully.

Manufactured Building Boards

There are a number of building boards made from wood veneers or the waste products of the timber industry that are convenient and economical materials to use in building construction. In general, they offer excellent bracing for the building frame, together with labour savings because they are available in large sizes requiring a minimum of fitting.

Some manufactured boards are designed with specific characteristics, such as fire resistance, ease of cleaning, high insulation value or resistance to weathering.

Plywood

Plywood is produced by gluing together three to seven veneers that have been peeled from logs. The grain of each successive veneer is angled at 90° from the previous one, resulting in a board that has considerable strength and rigidity in all directions. Waterproof glue is most commonly used, giving a product that is highly resistant to moisture. Waterproof glue panels should always be chosen for farm buildings. As the wood itself is not waterproof, the panels are still subject to swelling and shrinkage from moisture changes.

Grades of Plywood

Plywood is generally given four to five grades, based on the appearance of the surface veneers. Each panel has a double-letter grade to indicate the grade of the face of the panel and the back of the panel. The top-grade surface is generally free enough from defects to be finished naturally, while the second-best grade is good for painting. Lower grades are used for structural applications where appearance is of little importance. Theoretically, between 10 and 15 different grade combinations are possible. In practice, only a few tend to be available from timber merchants.

Sizes of Plywood Panel

As an example of a standard used in the region, the Kenya Bureau of Standards provides a standard with 12 panel sizes and 9 different thicknesses. Combining grades, panel sizes and thicknesses, there are numerous permutations, but only a few will be manufactured. The most common panel size is 2400 by 1200 mm, in thicknesses of 9 mm, 12 mm, 15 mm and 19 mm.

Plywood for Structural Members

Plywood panels are made from many different species of wood and have a wide range of strengths and stiffnesses. Specific strength characteristics for plywood can be provided by either the manufacturer or a trade association that publishes grade standards to which manufacturers adhere. In general, plywood panels should equal or exceed the strengths shown in table.

Table: Safe spans for plywood panels parallel to the grain of the plys.

Thickness	Load	
	167 Pa (170 kg/m^2)	4 790 Pa (490 kg/m^2)
9 mm	400 mm	-
12 mm	600 mm	-
15 mm	770 mm	300 mm
19 mm	925 mm	400 mm

Other Manufactured Boards

Blockboards and laminboards are made of strips of wood between 8 mm and 25 mm wide, glued together and covered with one or more veneers on each side. At least one pair of corresponding veneers will have the grain at right angles to the grain of the core. If the finish grain is to run parallel with the core, there must be at least two veneers per side.

The same 12 panel sizes listed for plywood are also listed for blockboard. However, the thicknesses are greater, ranging from 15 mm to 50 mm, in 5 mm increments. The same appearance, grades and types of glue listed for plywood also apply to blockboards. Blockboard panels are often used for doors.

Particle boards are formed by pressing chips or flakes of wood between pairs of heated platens so that the particles lie in random fashion with their longer dimensions parallel to the surface of the board. The chips are bonded with thermosetting synthetic resins. Depending on the size of the particles, these boards are variously known as particleboard, chipboard or waferboard. Strength and rigidity generally increase with density, but that alone is not a measure of quality, as moisture resistance varies considerably and most particleboards should not be used in moist locations.

Softboards are made from uncompressed woodchips or sugarcane fibres mixed with water and glue or resins, giving a density of less than 350 kg/m^3. They are inexpensive and can be used for wall or ceiling surfaces that are not subject to high-moisture conditions.

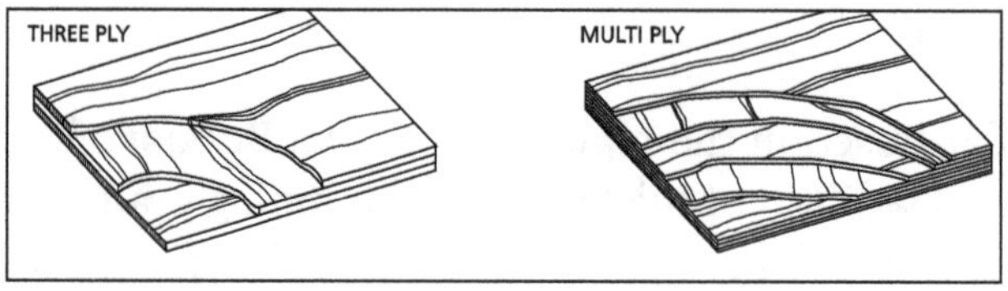

Plywood.

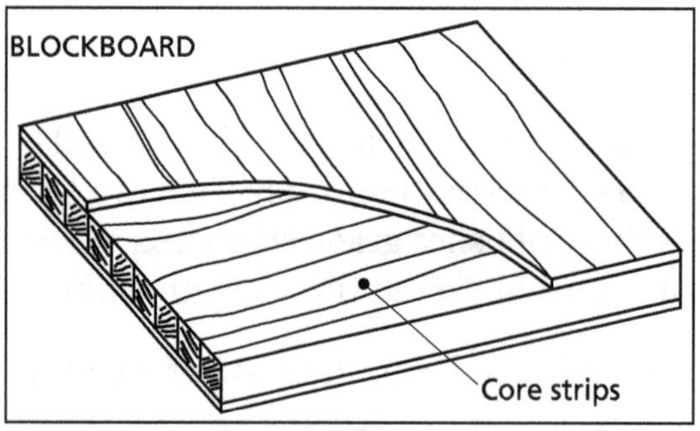

Blockboard.

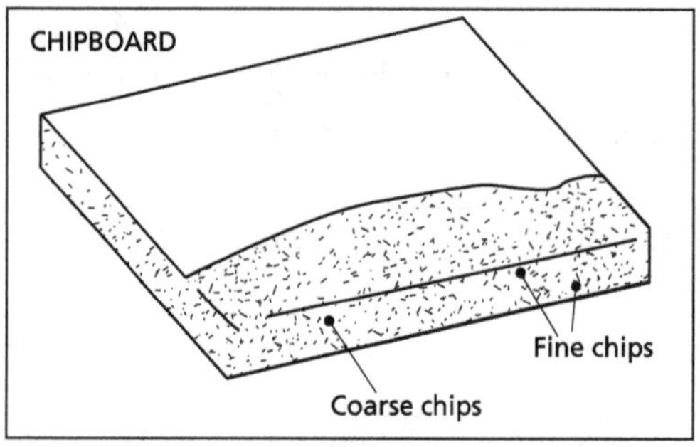

Particleboard.

Softboards have little resistance to rupture and must be supported frequently (300–400 mm) when installed. The 2400 mm by 1200 mm size is most common in thicknesses of 6.4 mm to 25 mm.

Mediumboards, with a density ranging from 350 kg/m^3 to 800 kg/m^3, are used for panelling, in particular those with a density at the higher end of the range. The most common size is 2400 mm by 1200 mm, and thicknesses range from 6.4 mm to 19.0 mm.

Hardboards are made of wood fibres compressed to more than 800 kg/m^3. They are usually smooth on one surface and textured on the other. The most common size is 2400 mm by 1200 mm size in thicknesses of 3–12.7 mm. An oil-treated grade labelled 'tempered' has good resistance to moisture.

Other Wood Products

Woodwool slabs consist of long wood shavings, mixed with cement, and formed into slabs 25–100 mm thick with a high proportion of thermal insulating voids. Although combustible, they are not easily ignited and provide good sound absorption.

Shingles are cut from clear rot-free timber logs. They are made about 2 mm thick at the top end and 10 mm thick at the bottom, and usually about 400 mm long. Some woods need treatment with preservatives before being used as roofing shingles, whereas others will last 10–15 years without treatment.

Sawdust is a by-product from sawmills. It is a good natural insulating material, and also a good bedding material for use in animal housing.

Wicker made from shrubs, bushes and trees is used either directly, for fencing or wall cladding, or sealed by smearing on mud, plaster, etc.

Other Organic Materials

Bamboo

Bamboo is a perennial grass with over 550 species, found in tropical, subtropical and temperate zones. It contains a large percentage of fibre, which has high- tensile, bending and straining capacity.

However, bamboos have some shortcomings that limit their application. The low durability of bamboo is one of its most serious defects, along with its flammability and tendency to split easily. This usually prevents the use of nails. Cutting a notch or a mortise in a bamboo drastically reduces its ultimate strength. The remedy is the use of nodes as places of support and joints, and the use of lashing materials (strings) in place of nails. Dry bamboo is extremely susceptible to fire, but it can be covered or treated with a fire-retardant material.

The strength properties of bamboo vary widely with species, growing conditions, position within the culm, seasoning and moisture content. Generally bamboo is as strong as timber in compression and very much stronger in tension. However, bamboo is weak in shear, with only about 8 percent of compressive strength, whereas timber normally has 20–30 percent. It is used mainly in building construction, for wall poles, frames, roof construction, roofing and water pipes and, after splitting, to form flattened boards or woven wall, floor and ceiling panels.

New stalks of bamboo are formed annually in clumps growing out of the spreading roots. The individual bamboo shoots complete their growth within a period of 4–6 months in the first growing season. A strengthening process takes place during the subsequent 2–3 years, and the culm reaches maturity after the fifth or sixth year, or even later depending on the species. It must be cut before blooming because it loses its resistance and dies after blooming. Some bamboos grow to 35 metres in height, while others are no more than shrubs. Diameters vary from 10 mm to 300 mm.

Bamboo without proper seasoning and preservative treatment will rot and be attacked by insects, particularly if used in moist locations, such as in earth foundations.

Bamboo Joints

As nailing causes splitting and notching, drastically reducing the strength of a bamboo culm, lashes are generally used as binding elements for framing. They may be split from the bamboo itself, or made from vines, reeds or the bark of certain trees. Soft galvanized wire is also used for binding. Bamboo can be kept from splitting when bending by boiling or steaming it first, then bending it while hot.

Lashing bamboo joints.

Several methods can be used for splitting bamboo culms. The edges of the strips can be razor sharp and should be handled carefully.

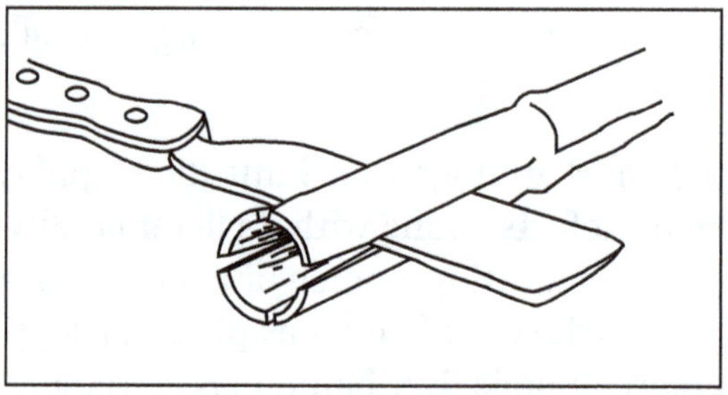

Make four cuts in the upper end of the culm with a splitting knife.

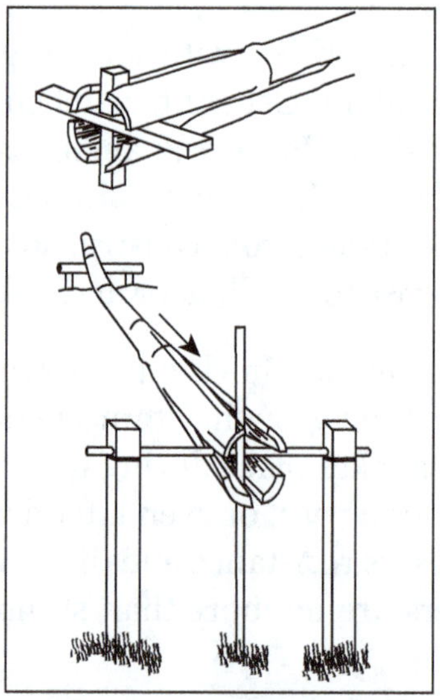

Split the culm the rest of the way by driving a hardwood cross along the cuts.

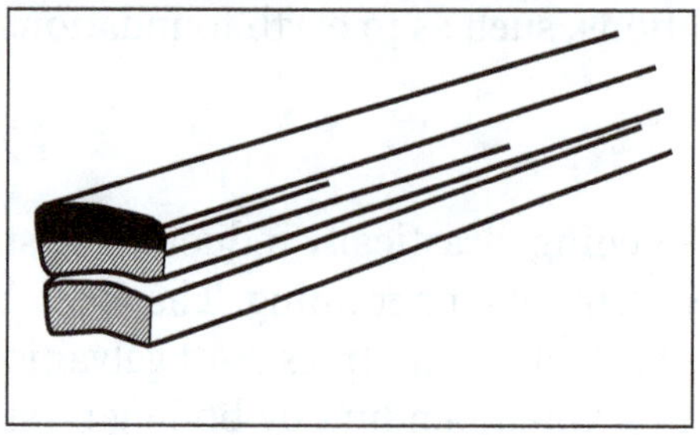

Use a knife to split the harder outer strip from the soft, pithy inner strip, which is usually discarded.

Bamboo Preservation

Immediately after cutting, the freshly cut lower end of the culm should be dusted with insecticide. The bamboo is then air seasoned for 4–8 weeks, depending on the ambient humidity. Bamboo should be stacked well off the ground so that air can circulate freely. When the culms have dried as much as conditions permit, they should be trimmed and all cut surfaces should be dusted with insecticide immediately. The seasoning is finished in a well-ventilated shelter where the culms are protected from rain and dew.

If the bamboo is to be stored for a long time, stacks and storage shelves should be treated with an insecticide every 6 months. Bamboo that has already been attacked by insects, fungus or rot should never be used for construction. Culms that have fissures, cracks or cuts in the surface should also be rejected.

Natural Fibres

Natural fibres have been used for building since ancient times. Fibrous materials can be used by themselves as roofing material or for walls and mats. Natural fibres can also be combined with hydraulic-setting binders to make various types of roofing board, wall board, block and shingle. Animal hair is often used for reinforcing plaster.

Thatch

Thatch, whether made of grass, reeds, palm or banana leaves, is susceptible to decay caused by fungi and insects, and to destruction by fire. Preservative treatment is desirable but expensive. A treatment combining copper sulphate, sodium chromate and acetic acid reduces attack by rot and may considerably increase the life span of a thatched roof.

Grass

The use of thatched roofs is common in many countries, and suitable grass can be found almost everywhere. When well laid and maintained, it can last for 10–20 years or longer.

A good-quality thatching grass must be fibrous and tough, with a minimum length of 1 metre. It should also have thin stems without hollows, a low content of easily digestible nutrients and the ability to withstand repeated wetting without decaying.

An annual treatment with a mixture of the following chemicals will improve the fire-resistance of a thatched roof, and also give some protection against decay: 14 kg ammonium sulphate, 7 kg ammonium carbonate, 3.5 kg borax, 3.5 kg boric acid, 7 kg alum and 200 kg water.

Reeds

Reeds must be dry before use as a building material, and can be impregnated or sprayed with copper-chrome preservatives to prevent rotting. Ammonium phosphate and ammonium sulphate are used to protect the reeds against fire.

Reeds can be woven into mats for use as wall or ceiling panels, shade roofs, etc. The mats can be plastered easily. In tropical areas, thatch from untreated reeds may last only 1 year but, if well laid, treated and maintained, it can last 5–10 years.

Sisal Stems

Before dying, at 7–12 years of age, the sisal plant forms a pole shoot to carry the flowers. The pole may reach a height of 6 metres or more and has a fibrous circumference, which makes it tough, but the inner parts are quite soft. Sisal poles have limited structural strength and durability, but are sometimes used for wall cladding in semi-open structures, such as maize cribs. The poles can be split and are joined in the same way as bamboo.

Sisal Fibre

Sisal fibre is one of the strongest natural fibres. It has traditionally been used as reinforcement in gypsum plaster sheets. Sisal fibres have the ability to withstand degradation from bacteriological attack better than other organic fibres, but are attacked by the alkalinity of cement. However, research has been carried out to make sisal fibre, like other natural fibre composites, into a reliable cement reinforcement for long-term use in exposed situations.

Coir Waste

Coir is a by-product of coconuts. The husk is used for making coir mats, cushions and as fuel. It can be mixed with cement, glue or resins, either to produce low-density boards with good insulating and sound- absorption properties, or to be compressed to make building boards. It is also used as reinforcement in cement for making roofing sheets.

Elephant Grass

Elephant grass is a tall plant similar to bamboo, but with the difference that the stem is not hollow. The fibres of the grass can be used to partly or wholly replace the asbestos in net and corrugated roofing sheets. However, the sheets are more brittle and have a slightly lower strength than asbestos-cement sheets.

Straw

Baled straw, if supported by a framework of wooden poles, can be used to construct temporary walls. Straw has also been used as raw material for manufactured building boards. Straw and split bamboo can be cement- plastered to permanent structures, such as vaults and domes, at low cost.

Natural Stone Products

Natural stones are strong in compression and are generally extremely durable, although deterioration may result from the action of soluble salts, wetting and drying, or thermal movement. According to the manner of their geological formation, all stones used in building fall into one of three classes: igneous, sedimentary or metamorphic.

Igneous rocks are mostly very hard and difficult to cut to size and shape. However, they are very durable.

Sedimentary rocks, such as sandstone and limestone, are used extensively for building. They are not difficult to work, yet are quite durable. Coral stone is found in coastal areas, where chips or small stones are used in mud walls. Coral stone is also cut into blocks and, although not very strong, can be used in foundations and walls in multistorey houses.

Metamorphic stones consist of older stones that have been subjected to intense heat and pressure, causing structural change. Thus clay becomes slate, limestone becomes marble and sandstone becomes quartzite. Slate develops cleavage planes during formation. Roofing slates are split along these planes. They make very durable roof surfaces, but require strong frames because of their weight.

At the building site, the stones can be dressed to obtain a smooth surface. Often only the side(s) that will be visible are dressed.

Stones may also be used in the forms and sizes in which they naturally occur, and be embedded in mortar for foundation and wall construction. Stones are also crushed and sorted for size and use. Small crushed stones are used in making concrete. Large sizes are used as hardcore for filling purposes.

Earth as a Building Material

Earth is one of the oldest materials used for building construction in rural areas. The advantages of earth as a building material are:

1. It is resistant to fire.
2. It is cheaper than most alternative wall materials, and is readily available at most building sites.
3. It has a very high thermal capacity, which enables it to keep the inside of a building cool when it is hot outside and vice versa.
4. It absorbs noise well.
5. It is easy to work using simple tools and skills.

These qualities encourage and facilitate self-help and community participation in house building.

Despite its good qualities, earth has the following drawbacks as a building material:

1. It has low resistance to water penetration, resulting in crumbling and structural failure.
2. It has a very high shrinkage/swelling ratio, resulting in major structural cracks when exposed to changing weather conditions.
3. It has low resistance to abrasion, and requires frequent repairs and maintenance when used in building construction.

However, there are several ways to overcome most of these weaknesses that make earth a suitable building material for many purposes.

Soil Classification

Soil and earth are synonymous when used in relation to building construction. The term 'soil' refers to subsoil, and should not be confused with the geological or agricultural definition of soil, which includes the weathered organic material in topsoil. Topsoil is generally removed before any engineering works are carried out, or before soil is excavated for use as a building material. Mud is the mixture of one or more types of soil with water.

There are several ways in which soil may be classified: by geological origin, by mineral content (chemical composition), by particle size or by consistency.

Particle Size

Soils are grouped and named according to their particle size.

Grading

The soil materials in table seldom occur separately, and this necessitates a further classification according to the percentage of each contained in the soil. This is shown in the soil classification triangle, which shows, for example, that a sandy clay loam is defined as soil that contains 50–80 percent sand, 0–30 percent silt and 20–30 percent clay.

Only a few mixes can be used successfully for building construction in the state in which they are found. However, many mixes can be improved to make good building material by correcting the mix and adding stabilizers.

The clay fraction is of major importance in earth construction because it binds the larger particles together. However, soils with more than 30 percent clay tend to have very high shrinkage/swelling ratios which, together with their tendency to absorb moisture, may result in major cracks in the end product. High-clay soils require very high proportions of stabilizer or a combination of stabilizers.

Some soils produce unpredictable results, caused by undesirable chemical reactions with the stabilizer. Black cotton soil, a very dark coloured clay, is an example of such a soil. Generally speaking, soils that are good for building construction purposes are characterized by good grading, i.e. they contain a mix of different-sized particles similar to the ratios in table where all voids between larger particles are filled by smaller ones. Depending on use, the maximum size of coarse particles should be 4–20 mm.

Laterite soils, which are widely distributed throughout the tropical and subtropical regions, generally give very good results, especially if stabilized with cement or lime. Laterite soils are best described as highly weathered tropical soils containing varying proportions of iron and aluminium oxides, which are present in the form of clay minerals, usually together with large amounts of quartz. Their colours range from ochre, through red, brown and violet to black. The darker the soil, the harder, heavier and more resistant it is to moisture. Some laterites harden on exposure to air.

Table: Classification of soil particles.

Material	Size of particles	Means of field identification
Gravel	60–2 mm	Coarse pieces of rock, which are round, flat or angular
Sand	2–0.06 mm	Sand breaks down completely when dry; the particles are visible to the naked eye and gritty to the touch.

Silt	0.06–0.002 mm	Particles are not visible to the naked eye, but slightly gritty to the touch. Moist lumps can be moulded but not rolled into threads. Dry lumps are fairly easy to powder.
Clay	Smaller than 0.002 mm	Smooth and greasy to the touch. Holds together when dry and is sticky when moist.
Organic	Up to several centimetres	Spongy or stringy appearance. The organic matter is fibrous, rotten or partially rotten, several centimetres deep, with an odour of wet, decaying wood.

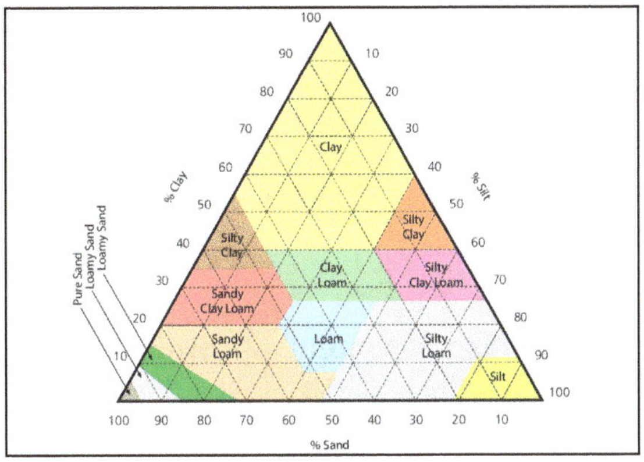

Soil-classification triangle.

Table: Soil gradings suitable for construction.

Use	Clay (%)	silt (%)	Clay& silt (%)	sand (%)	Gravel (%)	sand & Gravel (%)	Cobble (%)	organic matter (%)	soluble salts (%)
Rammed-earth walls	5–20	10–30	15–35	35–80	0–30	50–80	0–10	0–03	0–1.0
Pressed- soil blocks	5–25	15–35	20–40	40–80	0–20	60–80		0–03	0–1.0
Mud bricks (adobe)	10–30	10–40	20–50	50–80	-	50–80	-	0–0.3	0–1.0
Ideal, general-purpose mix	15	20	35	60	5	65	-	0	0

If the soil at hand is not suitable, it may be improved by adding clay or sand. The best soils for construction are sandy loam and sandy clay loam. Sandy clay gives fair results if stabilized.

Plasticity Index

Clays vary greatly in their physical and chemical characteristics. Although the extremely fine particles make it very difficult to investigate their properties, some can be conveniently expressed in terms of plasticity using standard tests.

Depending on the amount of moisture it contains, a soil may be liquid, plastic, semisolid or solid. As a soil dries, the moisture content decreases, as does the volume of the sample. With very high moisture content, the soil will flow under its own weight and is said to be liquid. At the liquid limit, the moisture content has fallen to the extent that the soil ceases to flow and becomes plastic; it is continuously deformed when a force is applied, but retains its new shape when the force is removed.

A further reduction of the moisture content will eventually cause the soil to crumble under load and not deform plastically. The moisture content at this point is known as the 'plastic limit'. The numerical difference between the moisture content at the liquid limit and at the plastic limit is called the 'plasticity index'. Both the liquid limit and the plasticity index are affected by the amount of clay and the type of clay minerals present.

A high liquid limit and plasticity index indicates a soil that has great affinity for water and will therefore be more susceptible to moisture movements, which can lead to cracks.

Example: The following index properties were determined for two soils X and Y.

Property	X	Y
Liquid limit	0.62	0.34
Plastic limit	0.26	0.19
Determine the plasticity index of X and Y.		

Solution

The plasticity index is the range of moisture content over which the soil remains plastic. The bigger this range, the greater the proportion of clay particles.

For soil X, plasticity index = liquid limit - plastic limit

= 0.62 - 0.26 = 0.36.

For soil Y, plasticity index = 0.34 - 0.19 = 0.15 Soil X contains more clay particles.

Soil-Testing Methods

Some soils are more suitable for building material than others. It is therefore essential to have a means of identifying different types of soil. There are a number of methods, ranging from laboratory tests to simple field tests. Laboratory soil tests are recommended for the production of buildings on a large scale (i.e. several houses).

As soils can vary widely within small areas, samples of the soil to be tested must be taken from exactly the area where soil is going to be dug for the construction. Soil samples should be collected from several places distributed over the whole of the selected area. First remove the topsoil (any dark soil with roots and plants in it), which is usually less than 60 cm. Then dig a pit to a depth of 1.5 metres, and collect soil for the sample at various depths between 0.8 metres and 1.5 metres. The total volume required for a simple field test is about a bucketful, whereas a complete laboratory test requires about 50 kg. Mix the sample thoroughly, dry it in the sun, break up any lumps and pass it through a 5–10 mm screen.

In the laboratory, the classification by particle size involves sieving the coarse-grained material (sand and gravel) and sedimentation for fine-grained material (silt and clay). The plasticity index is determined using the Atterberg limit test.

Soil tests will only give an indication of the suitability of the soil for construction purposes and the type and amount of stabilizer to be used. However, other properties, such as workability and behaviour during compaction, may eliminate an otherwise suitable soil. Soil tests should therefore be combined with tests on the finished products, at least where high strength and durability are required for the design and use.

For small projects, a simple sedimentation test combined with a bar shrinkage test normally gives enough information about the proportions of various particle sizes and the plastic properties of the soil.

Simple Sedimentation Test

This test gives an impression of the grading of the soil and allows the combined silt and clay content to be calculated. Take a large, clear glass bottle or jar with a flat bottom and fill it one-third full with soil from the sample. Add water until the bottle is two-thirds full. Two teaspoons of salt may be added to dissolve the soil more rapidly. Close the bottle, shake it vigorously, and allow the contents to settle for 1 hour. Shake it again and let it settle for at least 8 hours.

The soil sample should now show a fairly distinct line, below which the individual particles can be seen with the naked eye. Measure the thickness of the combined silt and clay layer above the line, and calculate it as a percentage of the total height of the soil sample.

The test tends to give a lower figure than laboratory tests, as a result of some silt and clay being trapped in the sand, and because some material remains suspended in the water above the sample.

The main disadvantage with this test is that the silt and clay fractions cannot be determined separately. As silt behaves differently from clay, this could result in mistaken conclusions about the soil's suitability for stabilization and as a building material.

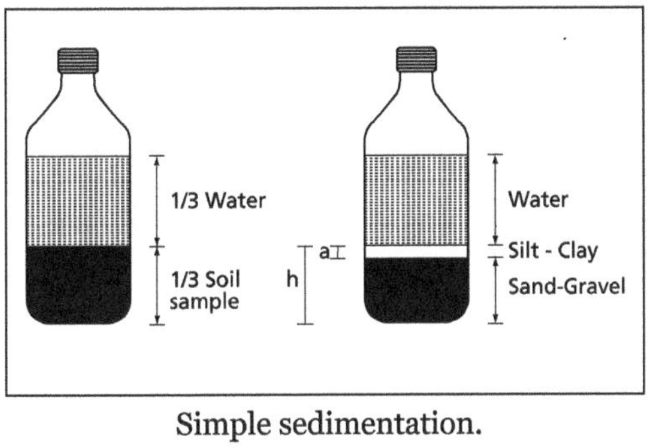

Simple sedimentation.

Bar Shrinkage Test

This test gives an indication of the plasticity index of the soil, because the shrinkage ratio of the soil when dried in its plastic state is related to its plasticity index.

A wooden or metal box without a top and with a square cross-section of 30–40 mm per side and a length of 500–600 mm, is filled with soil from the sample. Before filling, the soil should be mixed with water to slightly more than the liquid limit. The consistency is right when a V-shaped groove cut in the soil will close after about five taps on the box. Grease or oil the box, fill with the soil and compact it well, paying special attention to filling the corners. Smooth the surface by scraping off the excess soil. Place the box in the shade for 7 days. The drying can be hastened by placing the box at room temperature for 1 day, and then in an oven at 110 °C until the soil is dry.

If, after drying, the soil bar has more than three large cracks in addition to the end gaps the soil is not suitable. Measure the shrinkage ratio by pushing the dried sample to one end of the box and calculate the length of the gap as a percentage of the length of the box. The soil is not suitable for stabilization if the shrinkage ratio is more than 10 percent, i.e. a gap of 60 mm in a 600 mm long box. The higher the shrinkage ratio, the more stabilizer has to be used. The shrinkage ratio is calculated as follows:

$$\text{Shrinkage ratio} = \frac{(\text{Length of wet bar}) - (\text{Length of dried bar})}{\text{Length of wet bar}} \times 100$$

Soil Stabilization

The main weakness of earth as a building material is its low resistance to water. While overhanging eaves and verandas help considerably, tropical rains of any intensity can damage unprotected walls. Due to the clay fraction, which is necessary for cohesion, walls built of unstabilized soil will swell on taking up water, and shrink on drying. This may lead to severe cracking and difficulty in making protective renderings adhere to the wall.

However, the quality as a building material of nearly any inorganic soil can be improved considerably by the addition of a suitable amount of the correct stabilizer. The aim of soil stabilization is to increase the soil's resistance to destructive weather conditions, in one or more of the following ways:

- By cementing the soil particles together, leading to increased strength and cohesion.
- By reducing the movements (shrinkage and swelling) of the soil when its moisture content varies according to weather conditions.
- By making the soil waterproof or at least less permeable to moisture.

A great number of substances may be used for soil stabilization. Owing to the many different kinds of soil and types of stabilizer, there is no single solution for all cases. It is up to the builder to make trial blocks with various amounts and types of stabilizer.

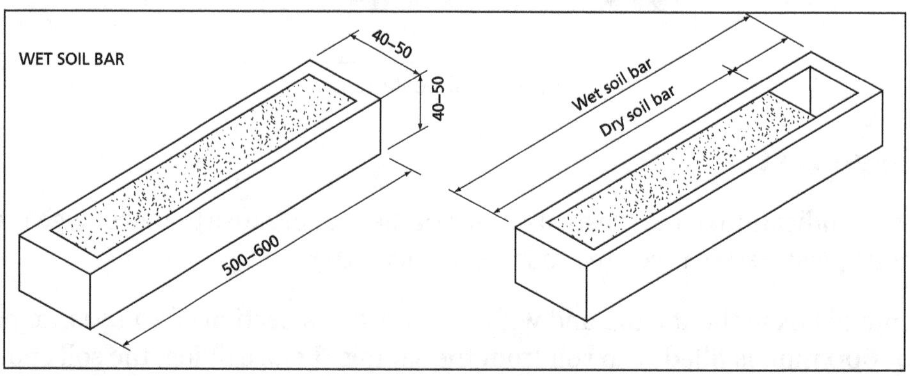

Box for bar shrinkage test.

Stabilizers in common use are:

- Sand or clay;
- Portland cement;
- Lime;
- Bitumen;

- Pozzolana (e.g. Fly ash, rice husk ash, volcanic ash);
- Natural fibres (e.g. Grass, straw, sisal, sawdust);
- Sodium silicate (water-glass);
- Commercial soil stabilizers (for roads);
- Resins;
- Whey;
- Molasses;
- Gypsum;
- Cow dung.

Many other substances may also be used for soil stabilization, although their use is not well documented and test results are scarce.

Sand or clay is added to improve the grading of a soil. Sand is added to soils that are too clayey, and clay is added to soils that are too sandy. The strength and cohesion of the sandy soil is increased, while moisture movement of a clay soil is reduced. Improved grading of the soil material does not stabilize the soil to a high degree, but will increase the effect, and reduce the required amount, of other stabilizers. The clay or clayey soil must be pulverized before mixing with the sandy soil or sand. This may prove difficult in many cases.

Portland cement greatly improves the soil's compressive strength and imperviousness, and may also reduce moisture movement, especially when used with sandy soils. As a rough guide, sandy soils need 5–10 percent cement for stabilization, silty soils, 10–12.5 percent and clayey soils, 12.5–15 percent. Compaction when ramming or pressing blocks will greatly improve the result.

The cement must be thoroughly mixed with dry soil. This can be rather difficult, especially if the soil is clayey. As soon as water is added, the cement starts reacting and the mix must therefore be used immediately (1–2 hours). If the soil-cement hardens before moulding, it must be discarded. Soil-cement blocks should be cured for at least 7 days under moist or damp conditions.

Non-hydraulic lime, or slaked lime, gives best results when used with fine soils, i.e. silty and clayey soils. Lime decreases moisture movement and permeability, by reaction with the clay, to form strong bonds between the soil particles. The amount of lime used varies between 4 percent and 14 percent. Lime breaks down lumps and makes it easier to mix clayey soils. Curing at high temperatures strengthens the cementing molecules, which should be an advantage in the tropics. The curing time is longer than for soil-cement.

A combination of lime and cement is used when a soil has too much clay for cement stabilization, or too little clay for an extensive reaction with the lime. Lime will make the soil easier to work and the cement will increase the strength. Equal parts of lime and cement are used. Mixing the dry soil with lime first makes the soil more workable. Blocks are cured for at least 7 days under moist conditions.

Bitumen (or asphalt) emulsion and cutback are used mainly to improve the impermeability of the soil and keep it from losing its strength when wet, but may cause some decrease in dry strength.

They are only used with very sandy soils because it would be very difficult to mix them with clayey soils. In its natural form, bitumen is too thick to be added to soil without heating, so it has to be thinned with other liquids to make it workable. The easiest way is to mix it with water to make an emulsion. After the emulsion has been added to the soil, the water will separate, leaving a bitumen film on the soil grains.

If the bitumen emulsion is fast-settling (i.e. the water separates too quickly before it is mixed into the soil), the bitumen must instead be dissolved in kerosene or naphtha. This mix is called 'cut-back' and should be handled with care because it represents a fire hazard and explosion risk. After a soil has been treated with cutback, it must be spread out to allow the kerosene to evaporate.

The bitumen content used is 2–4 percent, as any more may seriously reduce the compressive strength of the soil.

A combination of lime and pozzolana makes a binder that can be almost as good as portland cement. It is used in the same way as a combination of lime and cement, but 2-4 parts of pozzolana are mixed to 1 part of lime, and the curing time is longer than for ordinary cement.

Natural fibres, used in a mixing ratio of about 4 percent, greatly reduce moisture movement, but will make dry-soil blocks weaker and more permeable to water.

Sodium silicate, or water-glass, is best used to coat the outside of soil blocks as a waterproofing agent.

Cob

Cob is used extensively in tropical Africa, where suitable soils are obtainable over wide areas. The best soil mix consists of gravel, sand, silt and clay in roughly equal proportions. Sometimes chopped grass or straw is added to reduce cracking. If the clay content is high, sand may be added. Laterite makes an excellent material for cob walling.

When a suitable soil has been found, the topsoil is removed and the subsoil dug up. Water is slowly added to the loose soil, which is then kneaded by treading until the soil has a wet, plastic consistency. Natural fibres are added for stabilization if required.

The wet cob is rolled into balls or lumps measuring about 20 cm in diameter, which are then bedded on the wall to form courses about 60 cm high. The outside of the wall may be scraped smooth. In arid and semi-arid climates, this type of wall may last for years if built on a proper foundation and protected from rain by a roof overhang or veranda.

Wattle and Daub (Mud and Wattle)

This method of building small houses is very common in areas where bamboo or stalks (e.g. sisal) are available.

It consists of a framework of split bamboo, stalks or wooden sticks, supported by wood or bamboo poles. The soil, prepared as cob, is daubed on either side of the laths, which act as reinforcement. Although most soil is suitable for this construction, if it is too clayey there may be excessive cracking. To minimize cracking, stones are mixed with the soil, or laid in the wooden skeleton. When mudding the inside of a building, the soil is often taken from the floor. Although this increases the ceiling height, it greatly increases the likelihood of flooding during the rainy season.

During drying, the weight of the soil is transferred to the wooden structure, with the total weight of the construction eventually resting on the poles.

Wattle and daub construction generally has a short lifespan because of soil erosion, and the uneven settling of poles and damage by fungi and termites. However, the durability can be improved considerably (20–40 years) by using a proper foundation, raising the building off the ground, applying a surface treatment and using termite-resistant or treated poles.

Clay/Straw

The technique of building walls of clay/straw has been highly developed in China, where grain storage bins of up to 8 m in diameter, 8.5 m in height and a 250-tonne holding capacity have been constructed with these materials.

Any type of straw can be used, but it must be of good quality. The clay should be of strong plasticity, containing less than 5 percent sand. Some lime may be added for stabilization if the sand content is a bit too high.

First, the straw bundles are produced. The straw is pruned level at the root ends and then divided into two halves, which are turned in opposite directions and placed together so that they overlap by about two-thirds of the length of the straw. The straw bundle is then spread out flat and soaked with clay mud. Thorough covering of each straw is essential for the final strength. The straw is then twisted together, and any excess mud removed. The final clay/straw bundle should be thick in the middle and tapered at both ends, be of 80–100 cm in length and roughly 5 cm in diameter at the middle. The ideal proportion of straw and clay is 1:7 on a dry weight basis.

The clay/straw bundles are placed on the wall either straight and flat, or slightly twisted together. Walls for grain bins should have a thickness in centimetres equal to the internal diameter in metres + 12, i.e. a 6-metre diameter bin should have a wall thickness of at least 18 cm. It is important to compact the wall thoroughly during construction to ensure high density, strength and durability. The wall must be built in separate layers, usually about 20 cm, and be left to dry out to about 50 percent moisture content before the next layer is added.

Rammed Earth

This consists of ramming slightly damp soil between stout formwork using heavy rammers. It makes fairly strong and durable walls and floors when it is made thick enough with properly prepared, stabilized soil.

When used for walls, the soil may contain some cobble, but the maximum size should be less than one-quarter of the thickness of the wall. When cement is used for stabilization, it must be mixed with the dry soil by hand, or in a concrete mixer, until the dry mixture has a uniform greyish colour. The amount of cement required is approximately 5–7 percent for interior walls, 7–10 percent for foundations and exterior walls, and 10–15 percent for floors. However, the amount of stabilizer required will vary with the composition of the soil, the type of stabilizer and the use. For this reason, trial blocks should be made and tested to determine the correct amount of stabilizer.

Water is sprinkled on the soil while it is being mixed. If the soil is sticky because of high clay content, hand mixing will be necessary. When the correct amount of water has been added, the soil will form a firm lump when squeezed in the hand, and just enough moisture should appear on the surface to give a shiny appearance.

After the mixing has been completed, the soil should be placed in the formwork immediately. The formwork can be either fixed or sliding, but must be stout. The soil is placed in layers of about 10 cm, and each layer is thoroughly compressed with a ram weighing 8–10 kg before the next layer is placed. If water shows on the surface during ramming, the soil mix is too wet.

If cement or pozzolana has been used for stabilization, the product should be cured for 1–2 weeks in a moist condition before it is allowed to dry out. This can be done either by keeping the product enclosed in the formwork, or by covering it with damp bags or grass that are watered daily.

Adobe or Sun-Dried Soil (Mud) Blocks

The best soil for adobe can be moulded easily, when plastic, into an egg-size ball, and when it is allowed to dry in the sun it becomes hard, shows little deformity and no more than very fine cracks. If wide cracks develop, the soil does not contain enough silt or sand, and sand may be added as a stabilizer.

Preparing the Soil

When a suitable soil has been found, all topsoil must be removed. The soil is then loosened to a depth of l5 cm. If needed, water and sand are added and worked into the loose soil by treading it barefoot while turning the mass with a spade.

Water is added slowly and the soil mixed thoroughly until all lumps are broken up and it becomes homogeneous and plastic. When it is the right consistency for moulding, it is cast in a wooden mould made with one to three compartments and with dimensions as shown in figure.

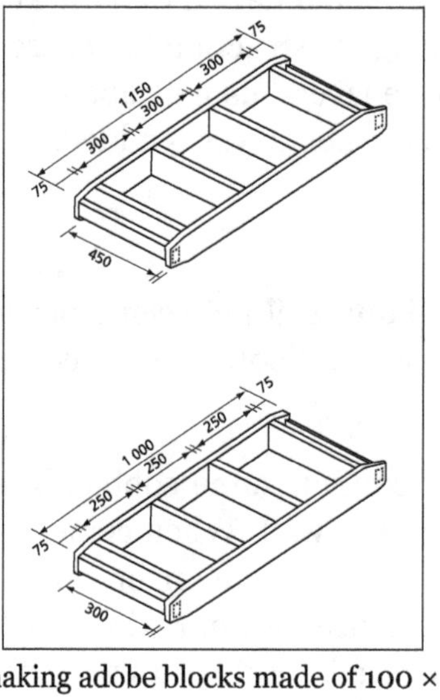

Wooden moulds for making adobe blocks made of 100 × 25 mm sawn timber.

Before the mould is used for the first time it should be thoroughly soaked in oil. Because of the shrinkage, the finished blocks will be smaller than the moulds and, depending on bonding, will give a wall thickness of about 230 mm, 270 mm or 410 mm.

Moulding the Blocks

To prevent sticking, the mould must be soaked in water before being placed on level ground, and filled with mud. The mud is kneaded until all corners of the mould are filled, and the excess is scraped off. The mould is lifted and the blocks are left on the ground to dry. The mould is dipped in water each time before repeating the process.

After drying for 3–4 days, the blocks will have hardened sufficiently to be handled, and are turned on edge to hasten drying. After a further 10 days, the blocks can be stacked loosely in a pile. Adobe blocks should dry out as slowly as possible to avoid cracks, with a total curing time of at least 1 month.

The quality of the blocks depends largely on the workmanship, especially the thoroughness with which they are moulded. If the quality is good, only 1 in 10 blocks should be lost from cracking, breakage or deformities.

Stabilized-Soil Blocks

When a suitable soil has been found, the topsoil should be removed and the subsoil dug out and spread out to dry in the sun for a few days.

Large particles and lumps must be removed before the soil is used, by breaking the larger lumps and passing all the soil through a 10 mm screen. If the proportion of gravel in the soil is high, a finer screen, of 4.5–6 mm, should be used. The wire screen, usually measuring about 1 metre square, is rocked in a horizontal position by one person holding handles at one end, with the other end suspended in ropes from above. The amount of loose, dry soil needed will normally be 1.4–1.7 times its volume in the compacted blocks.

Mixing

The amount of stabilizer to be used will depend on the type of soil, the type of stabilizer and the building component being made. Tables give a guide to the necessary minimum mixing ratio of soil-cement for blocks compacted in a mechanical press. For blocks compacted in a hydraulic press, the cement requirement can be reduced considerably, whereas slightly more will be needed for hand-rammed blocks. The correct proportion of stabilizer is determined by making test blocks with varying proportions of stabilizer, as described later.

Table: Cement/soil ratio related to shrinkage ratio in the bar shrinkage test.

Shrinkage	Cement to soil ratio
0–2.5%	1:18
2.5–5%	1:16
5–75%	1:14
7.5–10%	1:12

Table: Cement/soil ratio related to the combined silt and clay content in the simple sedimentation test.

Clay & silt content	Interior walls	Exterior walls	Foundations	Floor slab
0–10%	1:16	1:16	1:16	1:8
10–25%	1:22	1:16	1:16	1:11
25–40%	1:22	1:11	1:11	1:11

The importance of thoroughly mixing the dry soil first with the stabilizer and then with the moisture, in two distinct steps, cannot be emphasized too strongly.

The quantity of cement and dry soil is measured with a measuring box, bucket or tin (never with a shovel), and put either on a clean, even and hard surface for hand mixing, or into a drum-type mixer (concrete mixer). They are mixed until the dry mixture has a uniform greyish colour. Water is added, preferably through a sprinkler, while continuing the mixing. When the correct amount of water has been added, the soil, when squeezed into a ball, should retain its shape without soiling the hand. The ball should be capable of being pulled apart without disintegrating, but it should disintegrate when dropped from shoulder height on to a hard surface.

Table: Batching for stabilized-soil blocks.

			Number of blocks per 50 kg cement Size of blocks				
			290×140×50	290×140×90	290×140×120	290×140×140	290×215×140
1:22	5%	1 080 litres	366	203	152	130	85
1:18	6%	880 litres	301	167	125	107	70
1:16	7%	780 litres	268	149	111	95	62
1:14	8%	690 litres	235	131	98	84	54
1:12	9%	590 litres	203	113	84	72	47
1:11	10%	540 litres	187	104	78	66	43
1:10	11%	490 litres	170	94	71	61	39
1:9	12%	440 litres	154	85	64	55	36
1:8.5	13%	420 litres	146	81	61	52	34
1:8	15%	390 litres	138	76	57	49	32

Compaction by Hand-Ramming

Moulds with one or more compartments can be made from either hardwood or steel. The mould should have hinges at one or two corners to enable it to be opened easily without spoiling the block. As mould has no bottom, it is preferable to place it on a pallet, rather than directly on the ground, when moulding the block.

The mould is treated as often as required with oil to make the block surface smooth and to prevent the block from sticking to the mould. The soil mixture should be placed in layers in the mould, and each layer thoroughly compacted with a flat-bottomed ram weighing 4–5 kg. Each block may need as many as 80 good blows with the ram. The top of the block is levelled off, the block and mould are carried to the curing store where the mould is removed, then the whole process is repeated.

Compaction with a Mechanical Press

There are many mechanical block-making machines on the market, both motor-driven (enabling several blocks to be made at a time), and hand-operated.

They all consist of a metal mould in which a moist soil mix is compressed.

The moulding for a hand-operated press is carried out as follows:

1. The inside of the compaction chamber is cleaned and oiled, and a pallet is placed in the bottom, if required.
2. A measured amount of soil mix is poured into the compaction chamber and the soil is compacted into the corners by hand.

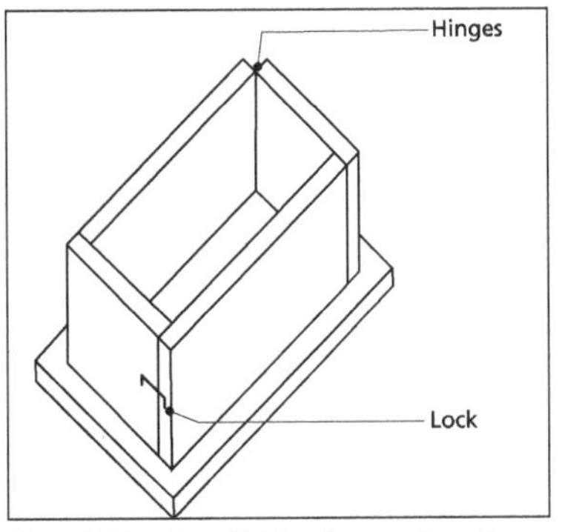

Mould for hand-rammed stabilized-soil blocks made of 20 mm planed timber.

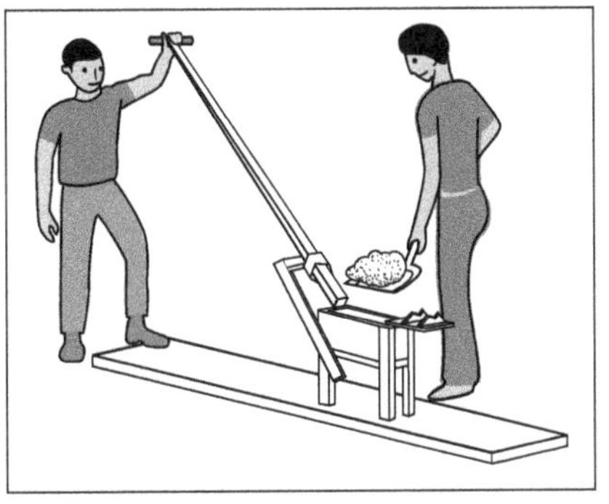

Mechanical press for block making.

3. The lid is closed and the handle pulled down. The amount of soil mix is correct if the handle can be moved down to stop slightly above a horizontal level.
4. The block is ejected and carried on the pallet to the curing site, before returning the pallet to the press for reuse.

Curing of Blocks

Soil-cement blocks should be placed on the ground, in the shade, as close together as possible, and be kept damp (e.g. with wet grass). After 1–2 days, the blocks can be carefully stacked and again kept damp for 1–2 weeks. After this period, the blocks are allowed to air dry for 2–3 weeks in a stacked pile before use.

Testing of Blocks

In the laboratory, dry strength and wet strength are determined by crushing two well-cured blocks in a hydraulic press: the first in a dry state, and the second after having been soaked in water for 24 hours. Durability is tested by spraying the blocks with water according to a standard procedure, and making observations for any erosion or pitting.

In order to find out how much stabilizer is required, the following simple weather-resistance test carried out in the field may give a satisfactory answer.

At least three different soil mixes with different stabilizer-soil ratios are prepared, and at least three blocks are made from each of the different mixes.

Mixing, compaction and curing must be done in the same way as for the block production process. At the end of the curing period, three blocks are selected from each set, immersed in a tank, pond or stream all night, and dried in the sun all day. This wetting and drying process is repeated for 7 days.

The correct amount of stabilizer to use is the smallest amount with which all three blocks in a set pass the test. While a few small holes are acceptable on the compaction surface, if many holes appear on all surfaces the blocks are too weak. If the blocks have passed the test and the dry block produces a metallic ring when tapped with a hammer, they will have satisfactory durability and hardness.

If the blocks fail the test, the reason may be any of the following:

- Unsuitable soil;
- Insufficient amount of stabilizer;
- Incorrect type of stabilizer;
- Inadequately dried or lumpy soil;
- Lumpy cement;
- Insufficient mixing of the stabilizer;
- Too much or too little water added;
- Not enough compaction;
- Incorrect curing.

Comparison of Masonry Units made of Various Materials

There are many methods for making bricks and blocks, several of which are suitable for local production because they are labour-intensive but do not require especially skilled labour.

The decision concerning which method of blockmaking or brickmaking to use depends on several factors, such as:

- The raw materials available;
- The characteristics of the soil;

- Raw material and production costs;
- The requisite standards of stability, compression strength, water resistance, etc. (3 n/mm^2 is often regarded as the minimum compressive strength for use in one-storey buildings);
- The existing facilities for the maintenance of production tools and machines;
- The required productivity.

Burnt-Clay Bricks

Burnt-clay bricks have good resistance to moisture, insects and erosion, and create a good room environment. They are moderate in cost and have medium to high compressive strength.

Bricks can be made using sophisticated factory methods, simple labour-intensive methods, or a range of intermediate mechanized technologies. Labour-intensive production methods are the most suitable for rural areas where the demand for bricks is limited. Handmade bricks will be of comparatively lower quality, especially in terms of compressive strength, and will tend to have irregular dimensions. However, they are economical and require little capital investment or transportation costs. Bricks made in this manner have been used in buildings that have lasted for centuries. Their longevity depends on the quality of the ingredients, the skill of the artisans and the climate in which they are used.

Brickmaking

Five main ingredients are required for brickmaking: suitable clay and sand, water, fuel and manpower. The clay must be easily available, plastic when mixed with small amounts of water, develop strength upon drying and develop hard and durable strength when burned.

Suitable soils contain 25–50 percent clay and silt and 50–75 percent coarser material, as determined by the simple sedimentation test. The soil must be well graded. Another test consists of hand-rolling moistened soil into a long cylinder 10 mm in diameter on a flat surface, and then picking it up by one end and letting it hang unsupported.

A soil is adequate for brickmaking if the piece of cylinder that breaks off is between 50 mm and 150 mm long. In the bar shrinkage test, using a mould 300 mm long and 50 mm wide and deep, a suitable soil should show no cracking, or only a little on the surface, and should shrink less than 7 percent, i.e. less than 20 mm.

The clay is obtained by chipping it out of a clay bank and, where necessary, mixing it with sand to form a mixture that will not crack during drying. Water is gradually added to make the clay plastic.

When making bricks, the mould must be cleaned periodically with water. Before each brick is formed, the mould is sprinkled with sand. A lump, or 'clot,' of clay that is only slightly larger than required for a brick is rolled into a wedge shape and then dipped in sand, before being thrown, point down, into the mould. When thrown correctly, the mould will be completely filled, and the excess clay is then shaved off the top with a bowcutter. The sand in the mould and on the clot helps to release the newly formed brick.

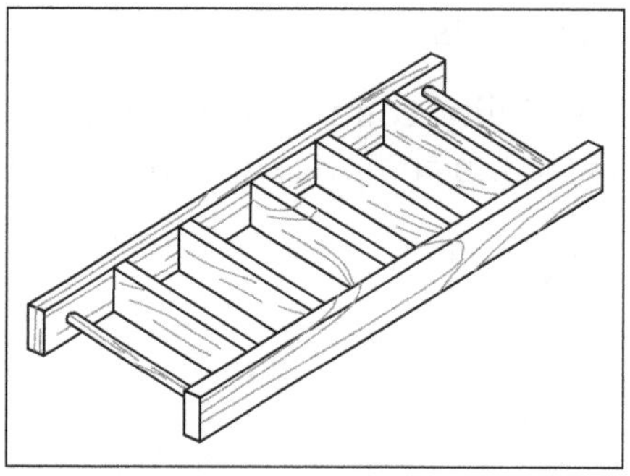

Mould for Brick-Forming.

The bricks should be left to dry for about 3 days in the place where they were made. They will then be strong enough to be stacked, as shown in figure, for at least 1 week of further drying. Clay tends to become lighter in colour when dry and, when sufficiently dried, the brick should show no colour variation throughout the section area when it is broken in half. During drying, the bricks should be protected from rain.

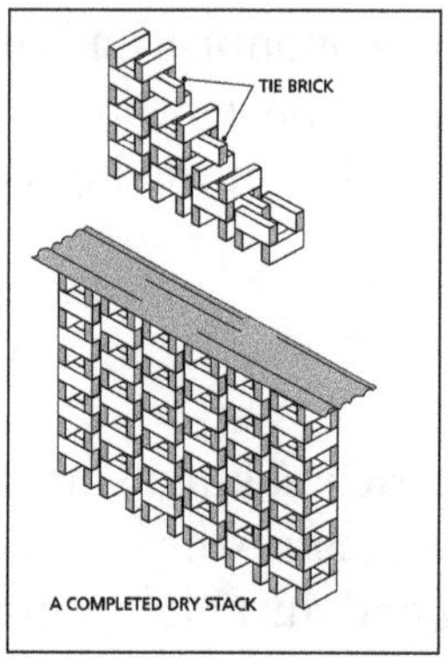

Stacking pattern for brick-drying.

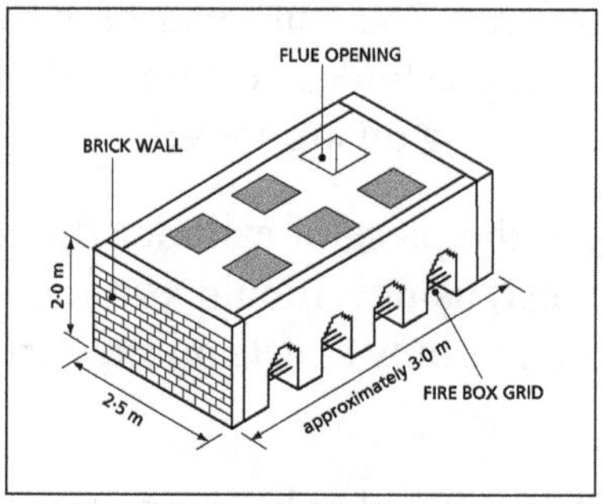

Kiln for brick-firing.

Kiln Construction and Brick-Firing

It is during firing that the bricks gain their strength. In high temperatures, the alkalis in the clay, together with small amounts of oxides of iron and other metals, chemically bond with the alumina and silica in the clay to form a dense and durable mass.

A kiln is a furnace or oven in which bricks are fired or heat-treated to develop hardness. Where brick-making is carried out on a large scale, the firing operation is performed in a continuous-process kiln, referred to as a 'tunnel' kiln. For brickmaking on a small scale, firing is a periodic operation where the bricks are placed in the kiln, the fire started and heat developed, and then, after several days of firing, the fuel is cut off from the fire and the entire kiln and its load are allowed to cool naturally. The kiln is filled with well-dried bricks, stacked in the same manner as during the drying process. The top of the stack in the kiln is then sealed with mud. Some openings are left for combustion gases to escape. Pieces of sheet metal are used to slide over the openings to control the rate at which the fire burns.

Although a range of fuels can be used in a kiln, wood or charcoal are the most common. When the kiln is at the prime heat for firing, a cherry-red hue develops (corresponding to a temperature range of 875–900 °C). This condition is maintained for about 6 hours. Sufficient fuel must be available when the burning starts because the entire load of bricks could be lost if the fires are allowed to die down during the operation. Firing with wood requires 4–5 days.

During firing, the bricks will shrink by as much as 10 percent. As they are taken out of the kiln, they should be sorted into different grades, the main criteria being strength, irregular dimensions, cracks and, sometimes, discoloration and stain.

Binders

When binders are mixed with sand, gravel and water, they make a strong and long-lasting mortar or concrete.

Binders can be broadly classified as non-hydraulic or hydraulic. The hydraulic binders harden through a chemical reaction with water, making them impervious to water and therefore able to harden under water. Portland cement, blast furnace cement (super sulphated), pozzolanas and high-alumina cement belong to the category of hydraulic binders. High-calcium limes (fat or pure limes) are non-hydraulic because they harden by reaction with the carbon dioxide in the air. However, if lime is produced from limestone containing clay, compounds similar to those in portland cement will be formed, i.e. hydraulic lime.

Lime

Non-hydraulic lime refers to high-calcium limes that are produced by burning fairly pure limestone (essentially calcium carbonate), in order to drive off the carbon dioxide, leaving calcium oxide or quicklime. The burning process requires a temperature of 900–1 100 °C. Quicklime must be handled with great care because it reacts with moisture on the skin and the heat produced may cause burns. When water is added to quicklime, considerable heat is generated, expansion takes place, breaking down the quicklime pieces into a fine powder, and the resulting product is calcium hydroxide, also called hydrated lime, or slaked lime.

After drying, the powder is passed through a 3-mm sieve, before being poured into bags for storage (in dry conditions) and distribution.

Process	Substance	Chemical formula
Burning	Limestone – quicklime	$CaCo_3$ - CaO + Co_2
Slaking	Quicklime – slaked lime	CaO + H_2o - $Ca(OH)_2$
Hardening	Slaked lime – limestone	$Ca(OH)$ + CO - $CaCo_3$ + H_2o

Slaked lime is used mainly in building because it is fat, i.e. it makes workable mortar and rendering and plaster mixes. Initially, a lime mortar becomes stiff by evaporation loss of water to absorptive materials such as bricks, but subsequent hardening depends on the chemical reaction with carbon dioxide from the air (carbonation) reforming the original calcium carbonate (limestone).

Non-hydraulic lime is also produced from limestone with a high content of magnesium carbonate. It is less easily slaked, but some of the remaining unslaked magnesium oxide may carbonate and produce greater strength than high-calcium lime.

Hydraulic lime is produced by mixing and grinding together limestone and clay material, and then burning it in a kiln.

It is stronger but less fat, or plastic, than non- hydraulic lime. During burning, the calcium oxide from the limestone reacts with siliceous matter from the clay to form dicalcium silicate. This compound may react with water, forming 'mineral glue' – tricalcium disilicate hydrate. The reaction is slow and may take weeks or months, but after some time a very good degree of strength is achieved.

The reaction that forms dicalcium silicate requires a very high temperature to be complete. In practical production, a lower temperature of 1 200 °C is used, leaving some of the ingredients in their original state. At this temperature, the limestone will lose the carbon dioxide and thus form quicklime. If the correct amount of water is added, the quicklime will slake, forming a fine powder. However, that excess water will lead to premature hardening caused by hydraulic reaction.

Cement

Portland cement hardens faster and develops considerably higher strength than hydraulic lime. This is because cement contains tricalcium silicate. However, the manufacturing process is much more complicated than for lime. The ingredients are mixed in definite and controlled proportions, before being ground to a very fine powder. The fine grinding is necessary because the formation of tricalcium silicate can only take place in a solid state, and therefore only the surface of the particles in the mix is accessible for the chemical reaction, which requires a temperature of 1 250–1 900 °C to be completed.

During burning, the small particles of limestone and clay are sintered together to form clinker. After cooling, the clinker is ground to cement powder, with a small amount of gypsum being added during the grinding. The finer the cement particles, the larger the surface area available for hydration by water, and the more rapidly setting and hardening occurs. Cement is normally sold in 50 kg bags, but occasionally is available in bulk at a lower price.

Ordinary portland cement is the least expensive, and by far the most widely used, type of cement. It is suitable for all normal purposes.

Rapid-hardening portland cement is more finely ground, which accelerates the chemical reaction with water and develops strength more rapidly. It has the same strength after 7 days as ordinary portland cement does after 28 days. Early hardening may be useful where early stripping of formwork and early loading of the structure is required.

Low-heat portland cement develops strength very slowly. It is used in very thick concrete work where the heat generated by the chemical reactions in ordinary portland cement would be excessive and lead to serious cracking.

Chemistry of Cement

The main components of standard portland cement are:

- Lime (calcium oxide: 66%) in the form of limestone;
- Silica (silicum dioxide: 22%), a component in most quartz, which forms the particles of clays;
- Aluminium oxide (4%), found in large quantities in many clays. The proportion of aluminium oxide in the clay can be adjusted by the addition of bauxite, which is mainly water-soluble aluminium oxide;
- Iron oxide (3%), found in iron ore and in clay;
- Magnesium oxide (2%);
- Sulphur dioxide (2%);
- Miscellaneous components (1%).

The manufacturing process aims to produce a material with a high content of tricalcium silicate, usually 55–62 percent of the crystals in the clinker. Other crystals formed are: about 15 percent dicalcium silicate (the same component as the hydraulic binder in hydraulic lime), 8–10 percent tricalcium aluminate and 9 percent tetracalcium aluminate ferrite. As cement sinters during burning, it is very important for no calcium oxide (quicklime) to remain in the finished product.

The quicklime will remain embedded in the clinker, even after very fine grinding, and will not be available for slaking until the hardening process of the cement is quite far advanced. When the quicklime particles are finally slaked, they expand and break the structure already developed. The proportion of limestone in the initial mix must therefore be no more than 0.1 percent.

When cement is mixed with water, it initiates the chemical reactions that are so important for hardening. The most important of these is the formation of tricalcium disilicate hydrate, 'mineral glue', from hydrated calcium oxide and silica.

$$2(3CaO\ SiO_2) + 6H_2O = 3CaO\ 2SiO_2\ 3H_2O + 3Ca(OH)_2$$

and

$$2(2CaO\ SiO_2) + 3H_2O = 3CaO\ 2SiO_2\ 3H_2O + Ca(OH)_2$$

The reaction between dicalcium silicate and water is slow and does not contribute to the strength of the concrete until a considerable time has elapsed. Aluminate would interfere with these processes, hence the addition of gypsum at the end of the manufacturing process. The gypsum forms an insoluble compound with the aluminate.

During the hydration process, the cement chemically binds water corresponding to about one-quarter of its weight. Additional water evaporates, leaving voids, which reduce the density, and therefore the strength and durability, of the end products.

Pozzolana

A pozzolana is a siliceous material which, in finely divided form, can react with lime in the presence of moisture at normal temperatures and pressures to form compounds possessing cementing properties. Unfortunately, the cementing properties of pozzolana mixtures are highly variable and unpredictable.

A wide variety of materials, both natural and artificial, may be pozzolanic. Silica constitutes more than half the weight of the pozzolana. Volcanic ash was the first pozzolana that the Romans used to make concrete for a host of large and durable buildings. Deposits of volcanic ash are likely to be found wherever there are active, or recently active, volcanoes.

Other natural sources of pozzolana are rock or earth in which the silica constituent contains the mineral opal, and the lateritic soils commonly found in Africa. Artificial pozzolana includes fly ash from the combustion of coal in thermoelectric power plants, burnt clays and shales, blast furnace slag formed in the process of iron manufacture, rice husk ash and the ash from other agricultural wastes.

The energy requirement for the manufacture of portland cement is very high. By comparison, lime and hydraulic lime can be produced with less than half the energy requirement, and natural pozzolana may be used directly without any processing. Artificial pozzolana requires some heating, but less than half that required for lime production.

Pozzolana and lime can be produced with much less sophisticated technology than portland cement. This means that pozzolana can be produced at relatively low cost and requires much less foreign exchange than cement. However, it takes 2–3 times the volume of pozzolana to make a concrete with the same strength as with portland cement, and this adds to the cost of transport and handling.

The main use of pozzolanas is for lime-pozzolana mortars, for blended pozzolanic cements and as an admixture in concrete. Replacing up to 30 percent of the portland cement with pozzolana will produce 65–95 percent of the strength of portland cement concrete at 28 days. The strength nominally improves with age because pozzolana reacts more slowly than cement, and at 1 year about the same strength is obtained.

Concrete

Concrete is a building material made by mixing cement paste (portland cement and water) with aggregate (sand and stone). The cement paste is the 'glue' that binds the particles in the aggregate together. The strength of the cement paste depends on the relative proportions of water and cement, with a more diluted paste being weaker. In addition, the relative proportions of cement paste and aggregate affect the strength, with a higher proportion of paste making stronger concrete.

The concrete hardens through the chemical reaction between water and cement, without the need for air. Once the initial set has taken place, concrete cures well under water. Strength is gained gradually, depending on the speed of the chemical reaction. Admixtures are sometimes included in the concrete mix to achieve certain properties. Reinforcement steel is used for added strength, particularly for tensile stresses.

Concrete is normally mixed at the building site and poured into formwork of the desired shape, in the position that the unit will occupy in the finished structure. Units can also be precast, either at the building site or at a factory.

Ingredients

Cement

Ordinary Portland cement is used for most farm structures. It is sold in paper bags containing 50 kg, or approximately 37 litres. Cement must be stored in a dry place, protected from ground moisture, and the storage period should not exceed a month or two. Even damp air can spoil cement. It should be the consistency of powder when used. If lumps have developed, the quality has decreased, although it can still be used if the lumps can be crushed between the fingers.

Aggregate

Aggregate or ballast is either gravel or crushed stone. Aggregates that pass through a 5 mm sieve are called fine aggregate or sand. and those retained are called coarse aggregate or stone. The aggregate should be hard, clean and free from salt and vegetable matter. Too much silt and organic matter makes the aggregate unsuitable for concrete.

To test for silt, place 80 mm of sand in a 200 mm-high transparent bottle. Add water up to a height of 160 mm. Shake the bottle vigorously and allow the contents to settle until the following day. If the silt layer, which will settle on top of the sand, is less than 6 mm, the sand can be used without further treatment. If the silt content is higher, the sand must be washed.

To test for organic matter, place 80 mm of sand in a 200 mm-high transparent bottle. Add a 3 percent solution of sodium hydroxide up to 120 mm. Note that sodium hydroxide, which can be bought from a chemist, is dangerous to the skin. Cork the bottle and shake it vigorously for 30 seconds, then leave it standing until the following day. If the liquid on top of the sand turns dark brown or coffee-coloured, the sand should not be used. A 'straw' colour is satisfactory for most jobs, but not for those requiring the greatest strength or water resistance. Note, however, that some ferrous compounds may react with the sodium hydroxide to cause the brown colour.

Grading of the aggregate refers to proportioning of the different sizes of the aggregate material and greatly influences the quality, permeability and workability of the concrete. With a well-graded aggregate, the various particle sizes intermesh, leaving a minimum volume of voids to be filled with the more costly cement paste. The particles also flow together readily, i.e. the aggregate is workable, enabling less water to be used. The grading is expressed as a percentage by weight of aggregate passing through various sieves. A well- graded aggregate will have a fairly even distribution of sizes.

Table: Suggested use for various concrete grades and nominal mixes.

Grade	Nominal mix	Use
C7 C10	1:3:8 1:4:6 1:3:6 1:4:5 1:3:5	Strip footings; trench fill foundations; stanchion bases; non-reinforced foundations; oversite concrete and bindings under slabs; floors with very light traffic; mass concrete, etc.

C15 C20	1:3:5 1:3:4 1:2:4 1:3:3	Foundation walls; basement walls; structural concrete; walls; reinforced floor slabs; floors for dairy and beef cattle, pigs and poultry; floors in grain and potato stores, hay barns, and machinery stores; septic tanks and water storage tanks; slabs for farmyard manure; roads, driveways, paving and walks; stairways.
C25 C30 C35	1:2:4 1:2:3 1:1.5:3 1:1:2	All concrete in milking parlours, dairies, silage silos, and feed and drinking troughs; floors subject to severe wear and weather conditions, or weak acid and alkali solutions; roads and paving in frequent use by heavy machinery and lorries; small bridges; retaining walls and dams; suspended floors, beams and lintels; floors used by heavy, small-wheeled equipment, such as lift trucks; fencing posts and precast concrete components.
C40 C50 C60		Concrete in very severe exposure; prefabricated structural elements; prestressed concrete.

The moisture content in sand is important because the sand mixing ratio often refers to kilograms of dry sand, and the maximum amount of water includes the moisture in the aggregate. The moisture content is determined by taking a representative sample of 1 kg. The sample is accurately weighed and spread thinly on a plate, soaked with spirit (alcohol) and burned while stirring. When the sample has cooled, it is weighed again. The weight loss amounts to the weight of the water that has evaporated, and is expressed as a percentage by dividing the weight lost by the weight of the dried sample. The normal moisture content of naturally moist sand is 2.5–5.5 percent. The amount of water added to the concrete mixture should be reduced by the same percentage.

Density is the weight per volume of the solid mass, excluding voids, and is determined by placing 1 kilo of dry aggregate in 1 litre of water. The density is the weight of the dry aggregate (1 kg) divided by the volume of water forced out of place. Normal values for density of aggregate (sand and stone) are 2 600–2 700 kg/m^3 and, for cement, 3 100 kg/m^3.

Bulk density is the weight per volume of the aggregate, including voids, and is determined by weighing 1 litre of the aggregate. Normal values for coarse aggregate are 1 500–1 650 kg/m^3. Although completely dry and very wet sand have the same volume, the bulking characteristic of damp sand gives it greater volume. The bulk density of typical naturally moist sand is 15–25 percent lower than coarse aggregate of the same material, i.e. 1 300–1 500 kg/m^3.

The size and texture of aggregate affects the concrete. The larger particles of coarse aggregate should not exceed one-quarter of the minimum thickness of the concrete member being cast. In reinforced concrete, the coarse aggregate must be able to pass between the reinforcement bars, 20 mm being generally regarded as the maximum size.

While aggregate with a larger surface area and rough texture, i.e. crushed stone, allows greater adhesive forces to develop, it will give less workable concrete.

Stockpiles of aggregate should be situated close to the mixing place. Sand and stone should be kept separate. If a hard surface is not available, the bottom of the pile should not be used in order to avoid defilement with soil. In hot, sunny climates, a shade should be provided, or the aggregate should be sprinkled with water to cool it. Hot aggregate materials make poor concrete.

Water

Water should be reasonably free of impurities such as suspended solids, organic matter and salts. This requirement is usually satisfied by using water which is fit for drinking. Sea water can be used if fresh water is not available, but not for reinforced concrete, as the strength of the concrete will be reduced by up to 15 percent.

Batching

The concrete mix should contain enough sand to fill all the voids between the coarse aggregate, enough cement paste to cover all particles with a complete film, and enough water to complete the chemical reaction. Requirements for batching ordinary concrete mixes of various grades and workability.

The water–cement ratio is an expression for the relative proportions of water, including the moisture in the damp aggregate, and cement in the cement paste. The strongest concrete is obtained with the lowest water-cement ratio which gives a workable mix that can be thoroughly compacted. Note that every 1 percent of water in excess of what is needed will reduce the strength by up to 5 percent. Water–cement ratio should however, not be below 0.4:1 since this is the minimum required to hydrate the cement.

The aggregate–cement ratio will influence on the concrete price since the amount of cement used per cubic meter will be changed. It is not possible to give a specific relationship between water-cement ratio, aggregate-cement ratio and workability, since it is influenced by the grading, shape and texture of the aggregate.

Workability describes the ease with which the concrete mix can be compacted. Workability can be increased by adding water to a given mixing ratio since this will increase the water–cement ratio and thereby reduce the strength. Instead it should be obtained by use of a well-graded aggregate (adjustment of the relative proportions of sand and stone), use of smooth and rounded rather than irregular shaped aggregate or by decreasing the aggregate–cement ratio.

Batching measuring is done by weight or by volume. Batching by weight is more exact but is only used at large construction sites. Batching by volume is used when constructing farm buildings. Accurate batching is more important for higher grades of concrete. Batching by weight is recommended for concrete of grade C30 and higher. Checking the bulk density of the aggregate will result in greater accuracy when grade C20 or higher is batched by volume. A 50 kg bag of cement can be split into halves by cutting across the middle of the top side of a bag lying flat on the floor. The bag is then grabbed at the middle and lifted so that the bag splits into two halves.

A bucket or box can be used as a measuring unit. The materials should be placed loosely in the measuring unit and not compacted. It is convenient to construct a cubic box with 335 mm sides, as it will contain 37 litres, which is the volume of 1 bag of cement. If the box is made without a bottom and placed on the mixing platform while being filled, it is easy to empty by simply lifting it. The ingredients should never be measured with a shovel or spade.

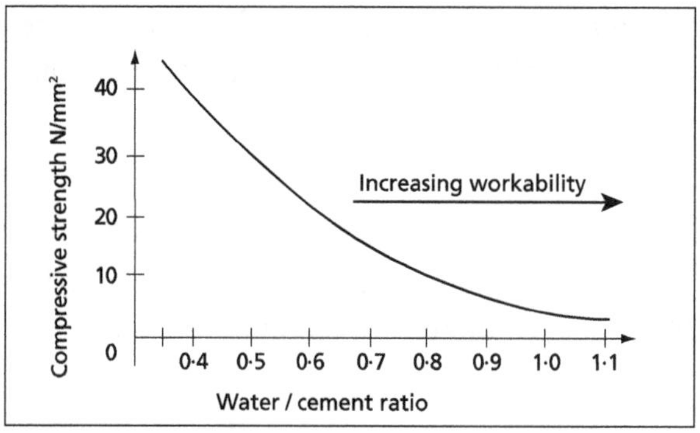

Relation between compressive strength and water/cement ratio.

Calculating of the amount of ingredients is done from the number of cubic metres of concrete required.

The sum of the ingredient volumes will be greater than the volume of concrete, because the sand will fill the voids between the coarse aggregate. The volume of the materials will normally be 30–50 percent greater than in the concrete mix – 5 percent to 10 percent is allowed for waste and spillage. Adding cement does not noticeably increase the volume. The above assumptions are used in example to estimate roughly the amount of ingredients needed. Example gives a more accurate method of calculating the amount of concrete obtained from the ingredients.

Table: Requirements per cubic metre for batching nominal concrete mixes.

		Naturally moist aggregate				Aggregate: cement	Sand to total aggregate
	Sand		Stones				
Proportions by volume	Cement (number of 50 kg bags	(m^3)	(tonnes)	(m^3)	(tonnes)	(ratio)	(%)
1:4:8	3.1	0.46	0.67	0.92	1.48	13.4	31
1:4:6	3.7	0.54	0.79	0.81	1.30	11.0	37
1 5:5	3.7	0.69	1.00	0.69	1.10	10.9	47
1:3:6	4.0	0.44	0.64	0.89	1.42	10.0	31
1:4:5	4.0	0.60	0.87	0.75	1.20	9.9	41
1:3:5	4.4	0.49	0.71	0.82	1.31	8.9	35
1:4:4	4.5	0.66	0.96	0.66	1.06	8.7	47
1:3:4	5.0	0.56	0.81	0.74	1.19	7.7	40
1:4:3	5.1	0.75	1.09	0.57	0.91	7.6	54
1:2:4	5.7	0.42	0.62	0.85	1.36	6.7	31
1:3:3	5.8	0.65	0.94	0.65	1.03	6.5	47
1:2:3	6.7	0.50	0.72	0.74	1.19	5.5	37
1:1:5:3	7.3	0.41	0.59	0.82	1.30	5.0	31
1:2:2	8.1	0.60	0.87	0.60	0.96	4.4	47
1:1:5:2	9.0	0.50	0.72	0.67	1.06	3.9	40
1:1:2	10.1	0.37	0.54	0.75	1.19	3.3	31

Example:

Calculate the amount of materials needed to construct a rectangular concrete floor measuring 7.5 m by 4.0 m and 7 cm thick. Use a nominal mix of 1:3:6. Fifty kilograms of cement is equal to 37 litres.

Total volume of concrete required = 7.5 m × 4.0 m × 0.07 m = 2.1 m^3

Total volume of ingredients, assuming 30 percent decrease in volume when mixed and 5 percent waste = 2.1 m^3 + 2.1 (30% + 5%) m^3 = 2.84 m^3

The volume of the ingredients is proportional to the number of parts in the nominal mix. In this case, there is a total of 10 parts (1+3+6) in the mix, but the cement does not affect the volume, so only the 9 parts of sand and stone are used.

Cement = (2.84 × 1) / 9 = 0.32 m^3 or 320

Sand = (2.84 × 3) / 9 = 0.95 m^3

Stone = (2.84 × 6) / 9 = 1.89 m^3

Number of bags of cement required =

320 / 37 = 8.6 bags, i.e. 9 bags have to be bought.

Weight of sand required =

0.95 m^3 × 1.45 tonnes / m^3 = 1.4 tonnes

Weight of stone required =

1.89 m^3 × 1.60 tonnes / m^3 = 3.024 tonnes

Maximum size of stones = 70 mm × 1 / 4 = 17.5 mm

Example:

Assume a 1:3:5 cement-sand-stone concrete mix by volume, using naturally moist aggregates and adding 62 litres of water. What will be the basic strength and volume of the mix if 2 bags of cement are used?

Additional assumptions:

Moisture content of sand: 4%

Moisture content of stones: 1.5% Bulk density of the sand: 1 400 kg/m^3

Bulk density of the stones: 1 600 kg/m^3

Solid density of aggregate materials: 2 650 kg/m^3 Solid density of cement: 3 100 kg/m^3

Density of water: 1 000 kg/m^3

1. Calculate the volume of the aggregate in the mix.

   Two bags of cement have a volume of 2 × 37 litres = 74 litres

   The volume of sand is 3 × 74 litres = 222 litres

   The volume of stones is 5 × 74 litres = 370 litres

2. Calculate the weight of the aggregates.

   Sand 222 / 1 000 m^3 × 1 400 kg/m^3 = 311 kg

   Stones 370 / 1 000 m^3 × 1 600 kg/m^3 = 592 kg

3. Calculate the amount of water contained in the aggregate

   Water in the sand is 311 kg × 4 / 100= 12 kg

   Water in the stones is 592 kg × 1.5 / 100= 9 kg

4. Adjust amounts in the batch for water content in aggregate.

   Cement 100 kg (unaltered)

   Sand 311 kg – 12 kg = 299 kg

   Stones 592 kg – 9 kg = 583 kg

   Total amount of dry aggregate = 299 kg + 583 kg = 882 kg

   Water = 62 kg + 12 kg + 9 kg = 83 kg

5. Calculate the water–cement ratio and the cement–aggregate ratio

   Water–cement ratio =

   (83 kg water) / 100 kg cement = 0.83

   Aggregate–cement ratio =

   (882 kg aggregate) / 100 kg cement = 8.8

   The water–cement ratio indicates that the mix has a basic strength corresponding to a C10 mix.

6. Calculate the 'solid volume' of the ingredients in the mix, excluding the air voids in the aggregate and cement.

   Cement 100 kg / 3 100 kg/m^3 = 0.032 m^3

   Aggregate 882 kg / 2 650 kg/m^3 = 0.333 m^3

   Water 83 kg / 1 000 kg / m^3 = 0.083 m^3

   Total = 0.448 m^3

The total volume of 1:3:5 mix obtained from 2 bags of cement is 0.45 m^3.

Note that the 0.45 m^3 of concrete is only two-thirds of the sum of the volumes of the components - 0.074 + 0.222 + 0.370.

Mixing

Mechanical mixing is the best way of mixing concrete. Batch mixers with a tilting drum for use on building sites are available in sizes of 85–400 litres. Power for drum rotation is supplied by a petrol engine or an electric motor, whereas the drum is tilted manually. The pear- shaped drum has internal blades for efficient mixing. Mixing should be continued for at least 2.5 minutes after all the ingredients have been added. For small- scale work in rural areas it may be difficult and rather expensive to use a mechanical mixer.

Table: Mixing-water requirements1 for dense concrete for different consistencies and maximum sizes of aggregate.

maximum size of aggregate	Water requirement (litres / m^3) for concrete		
	1/2–1/3 High workability	1/3–1/6 medium workability	1/6–1/2 Plastic consistency
10 mm	245	230	210
14 mm	230	215	200
20 mm	215	200	185
25 mm	200	190	175
40 mm	185	175	160

Batch mixer.

A simple hand-powered concrete mixer can be manufactured from an empty oil drum set in a frame of galvanized pipe. Figure shows a hand crank, but the drive can be converted easily to machine power.

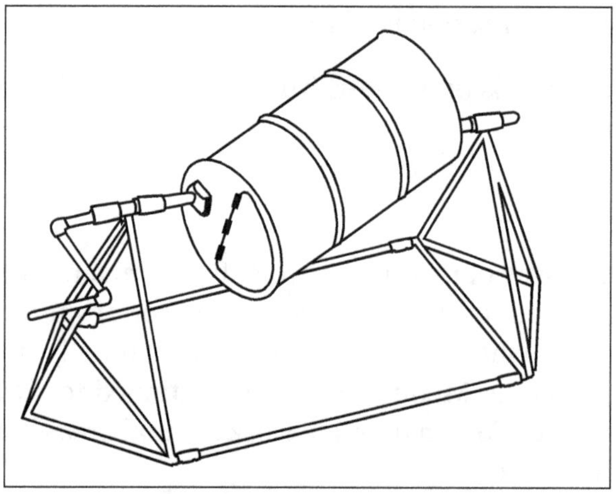

Home-built concrete mixer.

Hand mixing is normally used for small jobs. Mixing should be done on a close-boarded platform or a concrete floor near to where the concrete is to be placed, and never on bare ground because of the danger of earth contamination.

The following method is recommended for hand mixing:

1. The measured quantities of sand and cement are mixed by turning them over with a shovel at least 3 times.
2. About three-quarters of the water is added to the mixture a little at a time.
3. Mixing continues until the mixture becomes homogeneous and workable.
4. The measured quantity of stones, after being wetted with part of the remaining water, is spread over the mixture and mixing continues, with all ingredients being turned over at least 3 times during the process, using as little water as possible to obtain a workable mix.

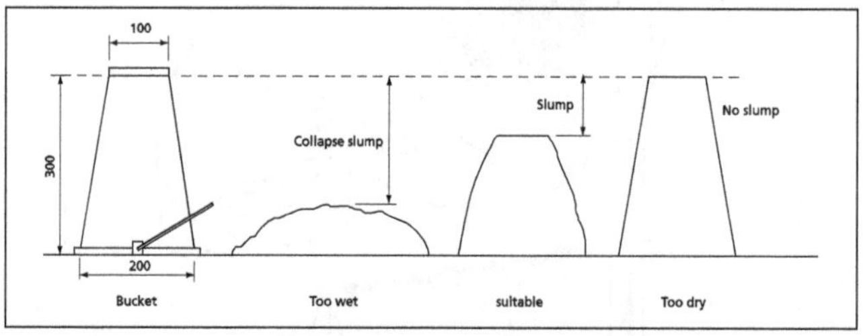

Concrete slump test.

All tools and the platform should be cleaned with water when there is a break in the mixing, and at the end of the day.

Slump Test

The slump test gives an approximate indication of the workability of the wet concrete mix. Fill a conically shaped bucket with the wet concrete mix and compact it thoroughly. Turn the bucket upside down on the mixing platform. Lift the bucket, place it next to the concrete heap and measure the slump, as shown in figure.

Placing and Compaction

Concrete should be placed with a minimum of delay after the mixing is completed, and certainly within 30 minutes. Special care should be taken when transporting wet mixes, because the vibrations of a moving wheelbarrow may cause the mix to segregate. The mix should not be allowed to flow, nor should it be dropped into position from a height of more than 1 metre. The concrete should be placed with a shovel in layers no deeper than 15 cm, and compacted before the next layer is placed.

When slabs are cast, the surface is levelled with a screed board, which is also used to compact the concrete mix as soon as it has been placed, to remove any trapped air. The less workable the mix, the more porous it is, and the more compaction is necessary. The concrete loses up to 5 percent of its strength for every 1 percent of entrapped air. However, excessive compaction of wet mixes brings fine particles to the top, resulting in a weak, dusty surface.

Manual compaction is commonly used for the construction of farm buildings. It can be used for mixes with high and medium workability, and for plastic mixes. Wet mixes used for walls are compacted by punting with a batten, stick or piece of reinforcement bar. Knocking on the formwork also helps. Less workable mixes, such as those used for floors and paving, are best compacted with a tamper.

Consistency	Slump	Use	Method of compaction
High workability	1/2–1/3	Constructions with narrow passages and/or complex shapes. Heavily reinforced concrete.	Manual
Medium workability	1/3–1/6	All normal uses. Non-reinforced and normally reinforced concrete.	Manual
Plastic	1/6–1/12	Open structures with fairly open reinforcement, which are heavily worked manually for compaction, such as floors and paving. Mass concrete.	Manual or mechanica
Stiff	0–1/2	Non-reinforced or sparsely reinforced open structures, such as floors and paving, which are mechanically vibrated. Factory prefabrication of concrete goods. Concrete blocks.	Mechanical
Damp	0	Factory prefabrication of concrete goods.	Mechanical or pressure

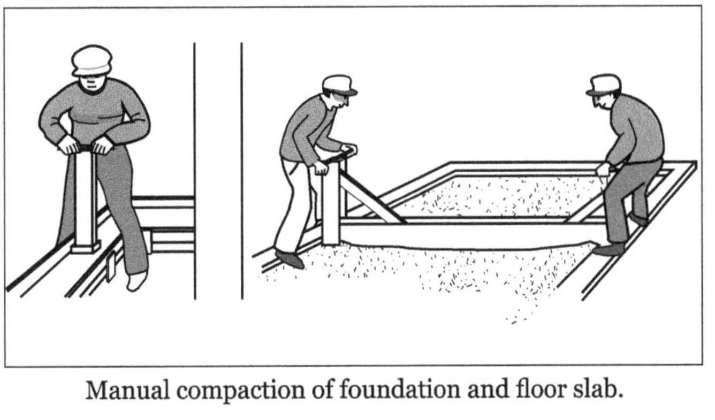

Manual compaction of foundation and floor slab.

Only mechanical vibrators are capable of compacting stiffer mixes thoroughly. For walls and foundations, a poker vibrator (a vibrating pole) is immersed in the poured concrete mix at points up to 50 cm apart. Floors and paving are vibrated with a beam vibrator.

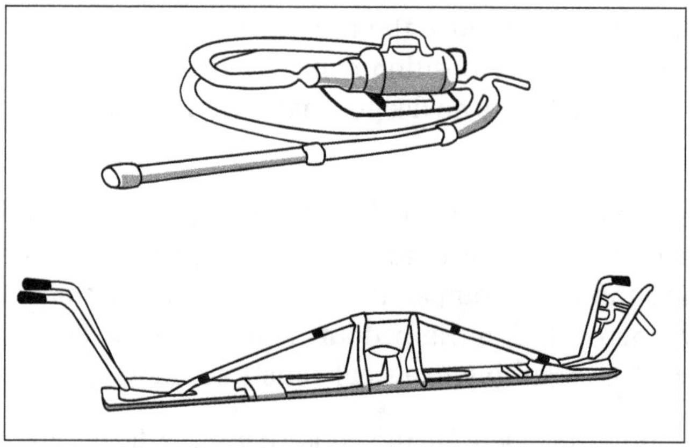

Mechanical vibrators.

Construction Joints

The casting should be planned in such a way that the work on a member can be completed before the end of the day. If cast concrete is left for more than 2 hours, it will set so much that there is no direct continuation between the old and new concrete. Joints are potentially weak and should be positioned where they will affect the strength of the member as little as possible. Joints should be straight, either vertical or horizontal. When resuming work, the old surface should be roughened and cleaned before being treated with a thick mixture of water and cement.

Formwork

Formwork provides the shape and surface texture of concrete members and supports the concrete during setting and hardening.

The simplest type of form is sufficient for pavement edges, floor slabs, pathways, etc.

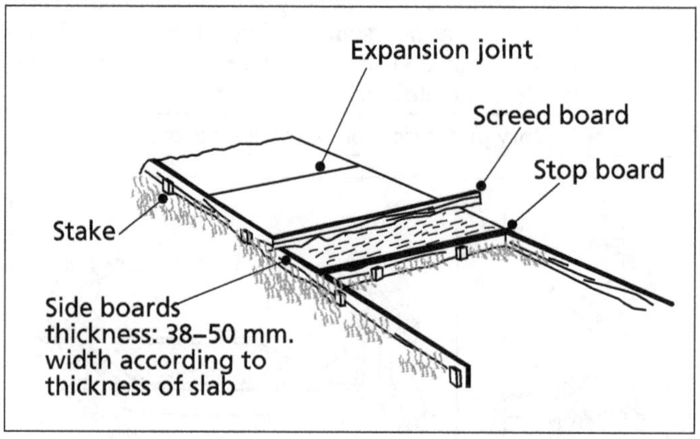

Simple type of formwork for a concrete slab.

In large concrete slabs, such as floors, cracks tend to occur early in the setting period. In a normal slab where watertightness is not essential, this can be controlled by laying the concrete in squares, with joints between them, allowing the concrete to move slightly without causing cracks in the slab. The distance between the joints should not exceed 3 metres. The simplest type is a called a dry joint. The concrete is poured directly against the already hardened concrete of another square.

A more sophisticated method is a filled joint. A minimum gap of 3 mm is left between the squares, and filled with bitumen or any comparable material.

Forms for walls must be strongly supported because when concrete is wet it exerts great pressure on the side boards. The greater the height, the greater the pressure. A concrete wall will not normally be thinner than 10 cm, or 15 cm in the case of reinforced concrete. If it is higher than 1 metre, it should not be less than 20 cm thick, to make it possible to compact the concrete properly with a tamper. The joints of the formwork must be tight enough to prevent loss of water and cement.

If the surface of the finished wall is to be visible, and no further treatment is anticipated, tongued and grooved boards, planed on the inside, can be used to provide a smooth and attractive surface. Alternatively, 12-mm plywood sheets can be used. The dimensions and spacing of studs and ties are shown in figure. The proper spacing and installation of the ties is important to prevent distortion or complete failure of the forms.

Not only must forms be well braced, they must also be anchored securely to prevent them from floating up, allowing the concrete to run out from underneath.

The forms should be brushed with oil and watered thoroughly before filling with concrete. This is done to prevent water in the concrete from being absorbed by the wooden boards, and to stop the concrete from sticking to the forms. Although soluble oil is best, used engine oil mixed with equal parts of diesel fuel is the easiest and cheapest material in practice.

If handled carefully, wooden forms can be used several times before they are abandoned. If there is a recurrent need for the same shape, it is advantageous to make the forms of steel sheets.

Although the formwork can be taken away after 3 days, leaving it for 7 days makes it easier to keep the concrete wet.

In order to save on material for the formwork and its supporting structure, tall silos and columns are cast using a slip form. The form is not built to the full height of the silo, and may in fact be only a few metres high. As casting of the concrete proceeds, the form is lifted. The work needs to proceed at a speed that allows the concrete to set before it leaves the bottom of the form. This technique requires complicated design calculations, skilled labour and supervision.

Curing Concrete

Concrete will set in 3 days, but the chemical reaction between water and cement continues for much longer. If the water disappears through evaporation, the chemical reaction will stop. It is therefore very important to keep the concrete wet (damp) for at least 7 days.

Premature drying out may also result in cracking caused by shrinkage. During curing, the strength and impermeability increases and the surface hardens against abrasion. Watering of the concrete should start as soon as the surface is hard enough to avoid damage, but not later than 10–12 hours after casting. Covering the concrete with sacks, grass, hessian, a layer of sand or polythene helps to retain the moisture and protects the surface from dry winds. This is particularly important in tropical climates.

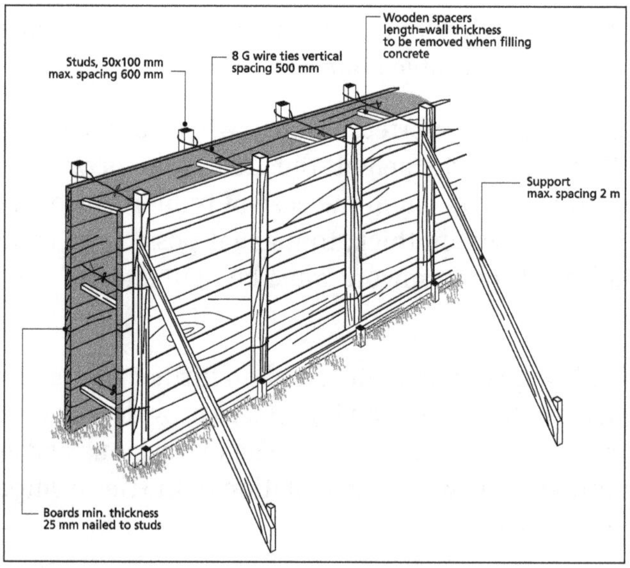

Dimensions and spacing of studs and ties in formwork for walls.

Temperature is also an important factor in curing. For temperatures above 0 °C and below 40 °C, strength development is a function of temperature and time. At temperatures above 40 °C, the stiffening and hardening may be faster than desired and result in lower strength.

Figure shows the approximate curing time needed to achieve characteristic compressive strength at various curing temperatures for concrete mixes using ordinary portland cement.

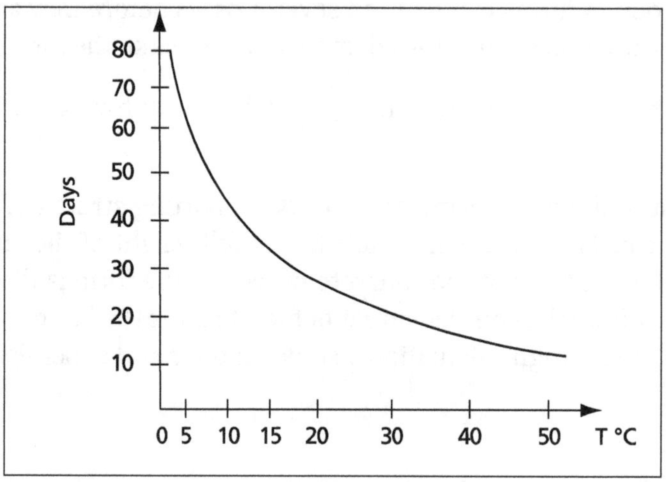

Curing times for concrete.

Finishes on Concrete

The surface of newly laid concrete should not be worked until some setting has taken place. The type of finish should be compatible with the intended use. In the case of a floor, a non-skid surface for humans and animals is desirable.

Tamped finish: The tamper leaves a coarse, rippled surface when it has been used to compact the concrete.

Tamper-drawn finish: A less pronounced ripple can be produced by moving a slightly tilted tamper on its tail end over the surface.

Broomed finish: A broom of medium stiffness is drawn over the freshly tamped surface to give a fairly rough texture.

Wood-float finish: For a smooth, sandy texture the concrete can be wood-floated after tamping. The float is used with a semicircular sweeping motion, the leading edge being slightly raised; this levels out the ripples and produces a surface with a fine, gritty texture, a finish often used for floors in animal houses.

Steel trowel finish: Steel trowelling after wood floating gives a smoother surface with very good wearing qualities. However, it can be slippery in wet conditions.

Surfaces with the aggregate exposed can be used for decorative purposes, but can also give a rough, durable surface on horizontal slabs. This surface can be obtained by removing cement and sand by spraying water on the new concrete, or by positioning aggregate by hand in the unset concrete.

Reinforced Concrete

Concrete is strong in compression but relatively weak in tension. The underside of a loaded beam, such as a lintel over a door, is in tension.

Stresses in a concrete lintel.

Concrete subject to tension loading must be reinforced with steel bars or mesh. The amount and type of reinforcement should be carefully calculated or, alternatively, a standard design obtained from a reliable source should be followed without deviating from the design.

Important factors affecting reinforced concrete:

1. The steel bars should be cleaned of rust and dirt before they are placed.
2. In order to obtain good adhesion between the concrete and the steel bars, the bars should be overlapped where they join by at least 40 times the diameter. When plain bars are used, the ends of the bars must be hooked.

3. The reinforcement bars should be tied together well and supported so that they will not move when concrete is placed and compacted.
4. The steel bars must be in the tensile zone and be covered with concrete to a thickness of 3 times the diameter, or by at least 25 mm, to protect them from water and air, which causes rusting.
5. The concrete must be well compacted around the bars.
6. Concrete should be at least C20 or 1:2:4 nominal mix, and have a maximum aggregate size of 20 mm.

Concrete floors are sometimes reinforced with welded steel mesh or chicken wire, placed 25 mm beneath the upper surface of the concrete, to limit the size of any cracking. However, such load-distributing reinforcement is necessary only when loadings are heavy, the underlying soil is not dependable, or when cracking must be minimized, as in water tanks.

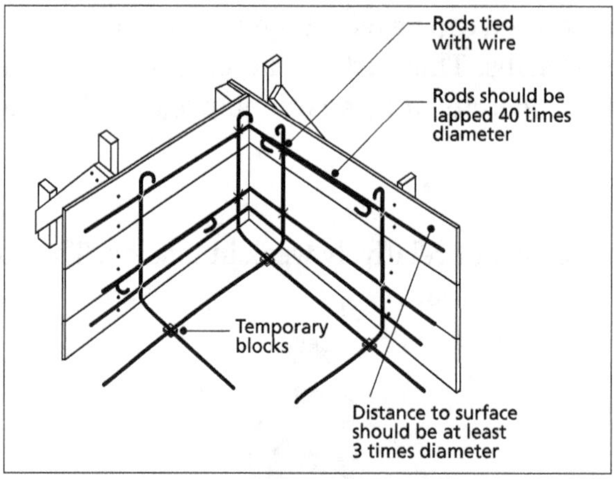

Placing reinforcement bars.

Concrete Blocks, Sand and Cement Blocks

It is faster to build with concrete blocks than with bricks, and using concrete blocks reduces the mortar requirement by half or more. If face-shell bedding is used, in which the mortar is placed only along the edges of the blocks, the consumption of mortar is reduced by a further 50 percent. However, the total amount of cement required for the blocks and mortar is far greater than that required for the mortar in a brick wall.

Concrete blocks are often made of 1:3:6 concrete with a maximum aggregate size of 10 mm, or a cement- sand mixture with a ratio of 1:7, 1:8 or 1:9. If properly cured, these mixtures produce concrete blocks with compression strength well above what is required in a one-storey building. The blocks may be solid, cellular or hollow. Cellular blocks have cavities with one end closed, while in hollow blocks the cavities pass through. Lightweight aggregate, such as cracked pumice stone, is sometimes used.

Blocks are made to a number of coordinating sizes, the actual sizes being about 10 mm less in order to allow for the thickness of the mortar.

Block Manufacturing

Blocks can be made using a simple block-making machine driven by an engine, or operated by hand. They can also be made using simple wooden moulds on a platform or floor. The mould can be lined with steel plates, to prevent damage during tamping and to reduce wear on the mould. Steel moulds are often used in large-scale production. The wooden mould is initially oiled overnight and need not be oiled each time it is filled. It is sufficient to wipe it clean with a cloth. The concrete, with a stiff or plastic consistency, is placed in the mould in layers, and each layer is compacted with a 3-kg rammer.

The mould in Figure has a lid made so that it can pass through the rest of the mould. The slightly tapered sides can be removed by lifting the handles, while holding down the lid with one foot.

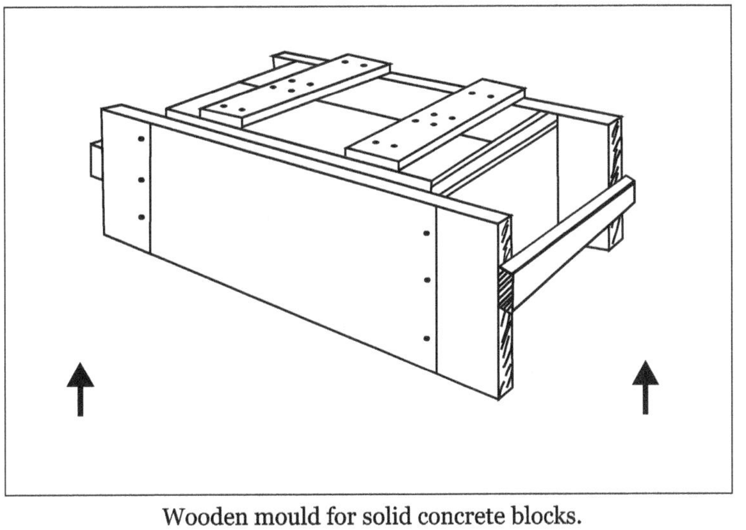

Wooden mould for solid concrete blocks.

The mould illustrated in Figure has a steel plate cut to the shape of the block, which is used as a lid and held down as the hollow-making pieces are withdrawn. Bolts are then loosened and the sides of the mould are removed with a swift motion. All parts of the mould should be slightly tapered so that they can be removed easily from the block.

As from the day after the blocks have been made, water is sprinkled on the blocks for 2 weeks during curing. After 48 hours, the blocks can be removed for stacking, but wetting must continue. After curing, the blocks are dried. If damp blocks are placed in a wall, they will shrink and cause cracks. To ensure maximum drying, the blocks are stacked interspaced, exposed to the prevailing wind and, in the case of hollow blocks, with the cavities laid horizontal to form a continuous passage for the circulating air.

Decorative and Ventilating Blocks

Decorative concrete or sand-cement blocks serve several purposes:

- To provide light and security without installing windows or shutters;
- To provide permanent ventilation;
- To give an attractive appearance.

In addition, some are designed to keep out rain, while others include mosquito proofing.

While blocks with a simple shape can be made in a wooden mould by inserting pieces of wood to obtain the desired shape, more complicated designs usually require a professionally made steel mould.

Mortar

Mortar is a plastic mixture of water and binding materials, used to join concrete blocks, bricks or other masonry units.

It is desirable for mortar to hold moisture, be plastic enough to stick to the trowel and the blocks or bricks, and to develop adequate strength without cracking.

Mortar need not be stronger than the units it joins. In fact, cracks are more likely to appear in the blocks or bricks if the mortar is excessively strong.

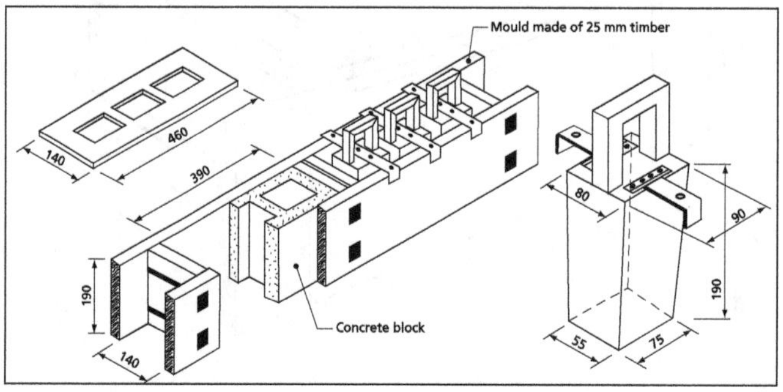

Mould for hollow or cellular concrete blocks.

There are several types of mortar, each suitable for particular applications and varying in cost. Most of these mortars include sand as an ingredient. In all cases, the sand should be clean, free of organic material, well graded (a variety of sizes) and not exceed 3 mm of silt in the sedimentation test. In most cases, particle size should not exceed 3 mm, as this would make the mortar 'harsh' and difficult to work with.

Lime mortar is typically mixed using 1 part lime to 3 parts sand. Two types of lime are available. Hydraulic lime hardens quickly and should be used within an hour. It is suitable for both above- and below-ground applications. Non-hydraulic lime requires air to harden, and can only be used above ground. If it is smoothed off while standing, a pile of this type of lime mortar can be stored for several days.

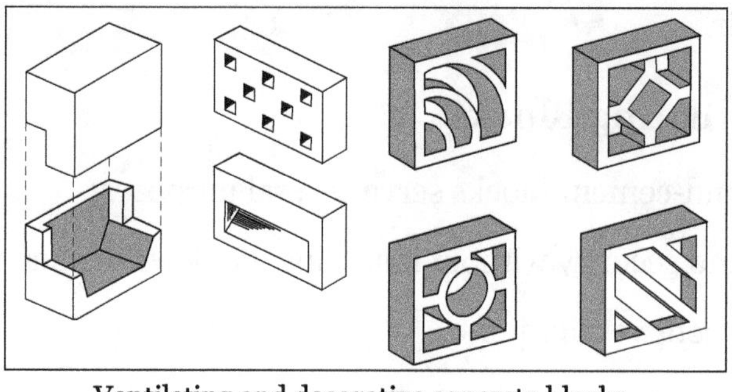

Ventilating and decorative concrete blocks.

Cement mortar is stronger and more waterproof than lime mortar, but it is difficult to work with because it is not 'fat' or plastic and falls away from the blocks or bricks during placement. In addition, cement mortar is more costly than other types. Consequently, it is used in only a few applications, such as a damp-proof course or in some limited areas where heavy loads are expected. A 1:3 mix using fine sand is usually required to obtain adequate plasticity.

Compo mortar is made with cement, lime and sand. In some localities, a 50:50 cement-lime mix is sold as mortar cement. The addition of the lime reduces the cost and improves workability. A 1:2:9, cement-lime-sand mix is suitable for general purposes, while a 1:1:6 is better for exposed surfaces, and a 1:3:12 can be used for interior walls, or stone walls, where the extra plasticity is helpful.

Mortar can also be made using pozzolana, bitumen, cutback or soil. A 1:2:9 lime-pozzolana-sand mortar is roughly equivalent to a 1:6 cement-sand mortar. Adobe and stabilized soil blocks are often laid in a mortar of the same composition as the blocks.

Tables and provide information on the materials required for a cubic metre of various mortars, and the amount of mortar per square metre, for several building units.

Starting with cement mortar, strength decreases with each type, although the ability to accommodate movement increases.

Table: Materials required per cubic metre of mortar.

Type	Cement bags	Lime (kg)	Sand (m^3)
Cement mortar 1:5	6.0	-	1.1
Compo mortar 1:1:6	5.0	100.0	1.1
Compo mortar 1:2:9	3.3	13.5	1.1
Compo mortar 1:8	3.7	-	1.1
Compo mortar 1:3:12	2.5	150.0	1.1
Lime mortar 1:3	-	200.0	1.1

Table: Mortar required for various types of wall.

Type of wall	Amount required per m^2 wall
11.5 cm brick wall	0.25 m^3
22.2 cm brick wall	0.51 m^3
10 cm sand-cement block wall	0.008 m^3
15 cm sand-cement block wall	0.011 m^3
20 cm sand-cement block wall	0.015 m^3

Finishing Mortar

This is sometimes used on floors and other surfaces to give a smooth finish, or as an extremely hard coating to increase resistance to wear. While such a top coating is prone to cracking, it seldom increases strength, and is difficult to apply without causing loose or weak parts. Concrete floors can normally be cast to the finished level directly, and be given a sufficiently smooth and hard surface without a top coating.

For coating, a mix of 1 part cement and 2–4 parts sand is used. The coating is placed in a 1–2 cm thick layer with a steel trowel. Before application, the surface of the underlying concrete slab should be cleaned and moistened.

Plastering and Rendering

The term 'plastering' is usually applied to interior walls and ceilings to give jointless, hygienic and usually smooth surfaces, often over uneven backgrounds. Exterior plastering is usually called 'exterior rendering'.

Cement plaster can be used on most types of wall, but it does not adhere well to soil-block walls, as shrinking and swelling tend to crack the plaster. The mixing ratio is 1 part cement and 5 parts sand and, if the plaster is too harsh, 0.5–1 part of lime can be added. The wall is first moistened and then the plaster is applied in two coats of about 5 mm each, allowing at least 24 hours between layers. Cement plaster should not be applied on a wall while it is exposed to the sun.

Dagga plaster is a mixture of clay soil (such as red or brown laterite), stabilizer and water. The plaster is improved by adding lime or cement as a stabilizer and bitumen for waterproofing. A good mixture is 1 part lime or cement, 3 parts clay, 6 parts sand, 0.2 parts bitumen and water. Dagga plaster is applied on previously moistened earth or adobe brick walls in a layer 10–25 mm thick.

Ferrocement

Ferrocement is a highly versatile form of reinforced concrete made with closely spaced light reinforcing rods or wire mesh, and a cement and sand mortar. It can be worked with relatively unskilled labour.

The function of the wire mesh and reinforcing rods is first to act as a lath, providing the form to support the mortar in its plastic state, while, in the hardened state, it absorbs the tensile stresses in the structure, which the mortar alone is not able to withstand.

The reinforcing can be assembled in any desired shape, and the mortar applied in layers to both sides. Simple shapes, such as water tanks, can be assembled using wooden sticks as support for the reinforcing while the first coat of mortar is applied.

The mortar should have a mixing ratio of 1:2 to 1:4 cement/sand by volume, using the richer mix for the thinnest structures. The water/cement ratio should be below 0.5/1.0. Lime can be added in the proportion 1 part lime to 5 parts cement in order to improve workability.

The mechanical behaviour of ferrocement depends on the type, quantity, orientation and strength of the mesh and reinforcing rods. The most common types of mesh used are illustrated in Figure.

Standard galvanized mesh (galvanized after weaving) is adequate. Although non-galvanized wire has adequate strength, the problem of rusting limits its use.

A construction similar to ferrocement has recently been developed for small water tanks, sheds, huts, etc. It consists of welded 150 mm-square reinforcement mesh (6 mm rods), covered with hessian and plastered in the same way as ferrocement.

Fibre-Reinforced Concrete

Fibre-reinforced concrete members can be made thinner than those with conventional reinforcement because there is no need for a corrosion-protection covering over the steel bars. The fibres improve flexible strength and resistance to cracking.

Commonly used fibres include asbestos, steel (0.25 mm diameter), sisal and elephant grass.

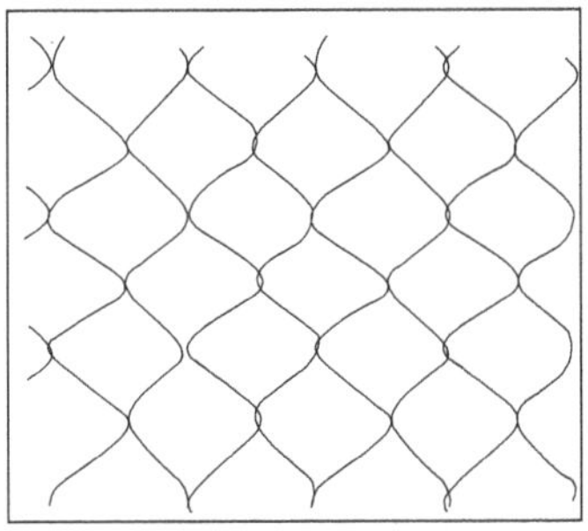

a. Hexagonal wire mesh (chicken wire mesh)

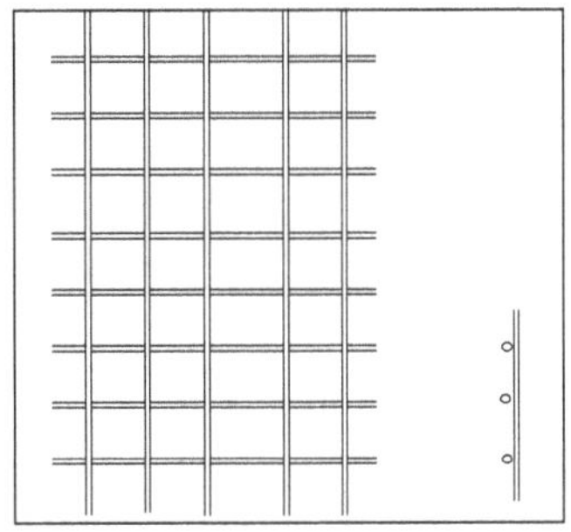

b. Welded wire mesh – strongest

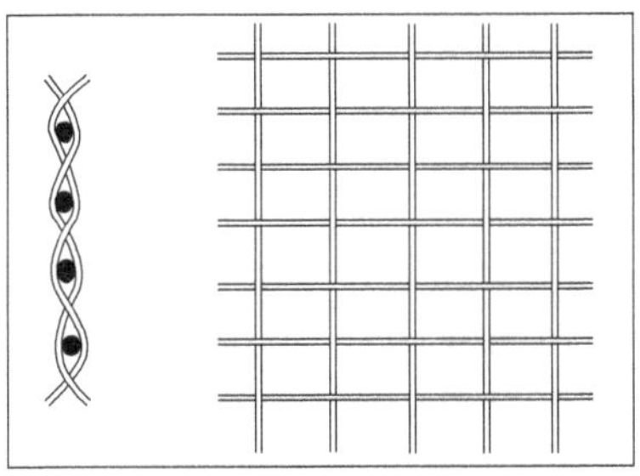

c. Woven mesh – strong

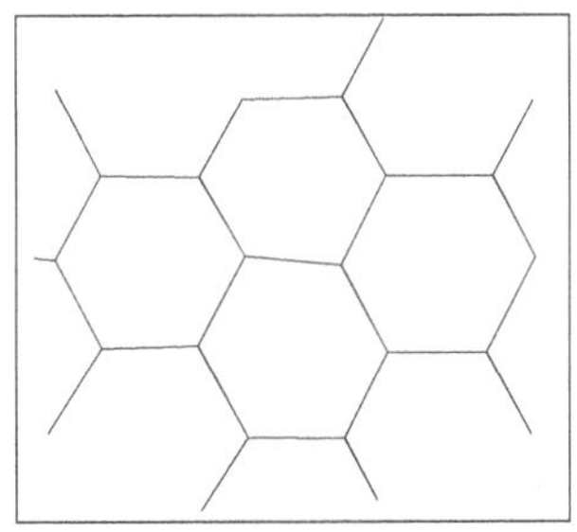

d. Expanded wire mesh

Reinforcement meshes for ferrocement.

Asbestos Cement

Asbestos, which is a silicate of magnesium, is found as a rock that can be split into extremely thin fibres ranging from 2 mm to 900 mm long. These have good resistance to alkalis, neutral salts and organic solvents, and the varieties used for building products have good resistance to acids. Asbestos is non-combustible and able to withstand high temperatures without alteration. Inhalation of asbestos dust causes asbestosis (a disease of the lungs) and asbestos is now used only where no alternative fibre is available. Workers must wear masks and take great care not to inhale any asbestos dust!

As the fibres are strong in tension and flexible, they are used as reinforcement with Portland cement, lime and bitumen binders, in asbestos-cement and asbestos- silica-lime products, vinyl floor tiles and in bitumen felts. Asbestos-cement is used in farm structures for corrugated roofing sheets, ridges and sanitary pipes.

Sisal-Fibre-Reinforced Cement

Sisal and other vegetable fibres have only recently come into use for cement reinforcement.

Sisal fibre can be used as short, discontinuous fibres (15–75 mm in length), or as continuous long fibres exceeding 75 mm in length. Sometimes both short and long fibres are used together. The manner in which the fibres are incorporated into the matrix affects the properties of the composite, both in the fresh state and in the hardened state.

Sisal fibres may deteriorate if not treated. Although the alkalinity of the concrete helps to protect the fibres from outside attack, it may itself attack the fibres chemically by decomposing the lignin.

Sisal fibre reinforcing is used with various cement- sand mixing ratios, depending on the use:

- Wall plastering - 1:3
- Guttering - 1:2
- Roofing tiles - 1:1
- Corrugated roofing sheets - 1:0.5

The sand should be passed through a sieve with 1.5 mm to 2 mm holes (e.g. mosquito netting). The mixing water must be pure and the mix kept as dry as possible, while still being workable.

Between 16 grams and 17 grams of short (25 mm), dry sisal fibres are added to the mix for each kilogram of cement. The short fibres are mixed into the dry cement and sand before adding water. As sisal fibres have a high water-absorption capacity, some extra water may have to be added to the mix to compensate for this.

When mixing, there is a tendency for the fibres to ball and separate out from the rest of the mix. This tendency will increase with longer fibres but, if fibres shorter than 25 mm are used, the reinforcing effect will be reduced. In most cases, the mix is then trowelled onto a mesh of full-length sisal fibres.

Making Corrugated Reinforced Roofing Sheets

Home-made reinforced corrugated roofing is usually cast to standard width, but to only 1 metre in length because of its additional weight. Commercial asbestos- cement roofing is heavier than corrugated steel, and the home-made sheets are still heavier. Special attention must therefore be given to rafter or truss sizes to ensure a safe structure.

Although the casting procedure for sisal-fibre- reinforced cement is tricky, once the proper equipment has been assembled and several sheets have been made, the process becomes much easier.

A concrete block cast over a 1-metre length of asbestos-cement roofing is needed as a face for casting the roof sheets. The block is cast within a 100 mm-high form, which will give a block of sufficient strength after a few days curing. Two or more 1-metre lengths of asbestos-cement roofing will be needed, as well as a piece of 18 mm plywood, measuring 1.2 metres by 1.2 metres, and a sheet of heavy-duty polythene, measuring 2.25 metres long and 1 metre wide. The polythene is folded in the middle and a thin batten, measuring 9 mm by 15 mm, is stapled at the fold. Strips of 9 mm plywood or wood are nailed along two edges of the plywood sheet, leaving exactly 1 metre between them, as shown in figure.

Below are the steps to follow in the casting procedure:

- Fit an asbestos cement sheet onto the moulding block and cover with the piece of plywood, with the edge strips at the ends of the sheet. The polythene is placed over the plywood and the top sheet is folded back off the plywood.

- Prepare a mix of 9 kg cement, 4.5 kg sand, 150 grams of short sisal fibres (25 mm) and 4.5 litres of water. Also prepare four 60-gram bundles of sisal fibres that are as long as possible.
- Use one-third of the mortar mix to trowel a thin, even layer over the polythene. Take two of the four sisal bundles and distribute the fibres evenly, with the second bundle at right angles to the first, forming a mat of fibres. This is covered with mortar and another mat made from the remaining two bundles. Finally, all the sisal is covered with the remaining mortar, and the surface is screeded even with the edge strips on the plywood.
- Cover with the top sheet of polythene, ensuring that the mortar is of even thickness all over and that no air bubbles remain under the polythene.
- While holding the batten strip at the fold in the polythene, carefully remove the plywood sheet to allow the new sisal-cement sheet to fall onto the asbestos-cement sheet. At the same time, press the new sheet into the corrugations using a PVC drain pipe 90 mm in diameter. Compact the new sheet by placing another asbestos sheet on top, and treading on it. Holes for mounting are punched with a 5 mm dowel 25 mm from the end in the gulleys (crests when mounted on the roof) of the fresh sheet.
- Remove from the moulding block the asbestos sheet bearing the sisal-cement sheet, and leave it until the cement in the new sheet has set (preferably 2 days). Then carefully remove the new sheet, peel off the polythene and cure the new sheet for at least 1 week, preferably immersed in a water tank.
- If more polythene and asbestos-cement sheets are available, casting can proceed immediately.

Walls using the Sisal-Cement Plastering Technique

Soil blocks can be used for inexpensive walls with good thermal insulation. However, they are easily damaged by impact and eroded by rain. One way of solving these problems is to plaster the face of the wall. Ordinarily, mortar plaster tends to crack and peel off, as it does not expand at the same rate as the soil. This can be overcome by letting long sisal fibres pass through the wall, to be incorporated into the mortar on each face. The double skin so formed provides sufficient strength and waterproofing to the wall to enable soil blocks to be laid without mortar between the blocks to join them.

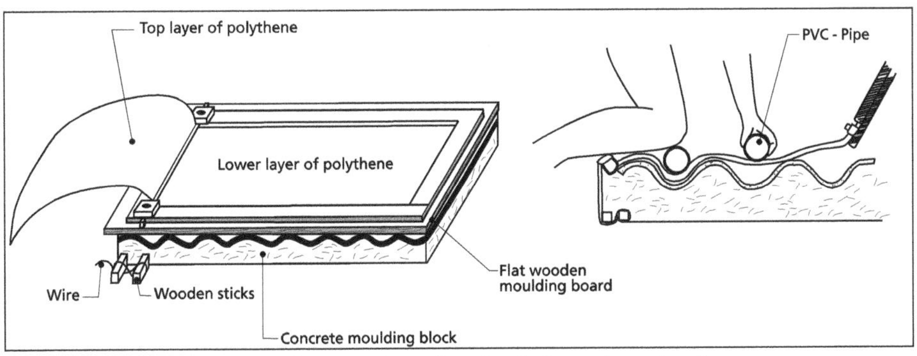

Plywood casting board and polythene 'envelope'.

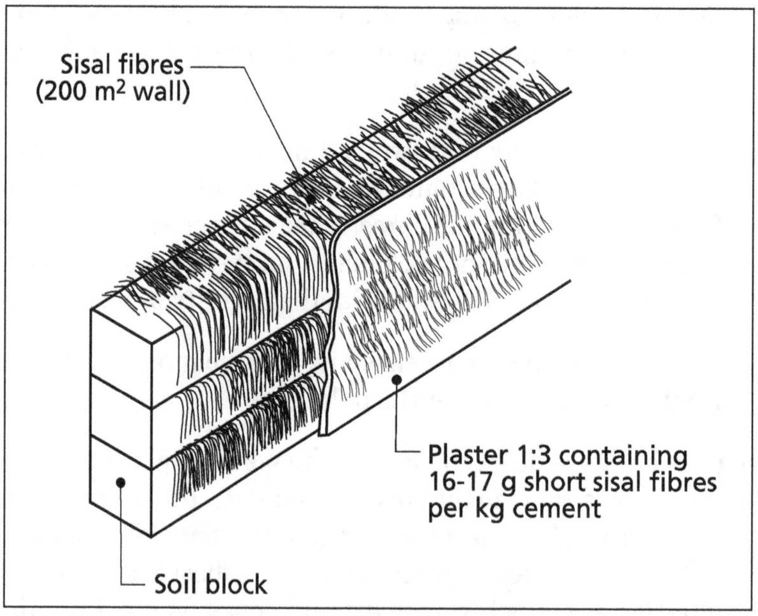

Sisal-cement plastering technique.

Metals

Several ferrous metals (those containing iron) are useful in the construction of farm buildings and other rural structures. Cast iron is used for making sanitary waste pipes and fittings.

Steel consists of iron, plus a small percentage of carbon in chemical combination. High-carbon or 'hard' steel is used for tools with cutting edges. Medium- carbon steel is used for structural members such as I-beams, reinforcing bars and implement frames. Low- carbon or 'mild' steel is used for pipes, nails, screws, wire, screening, fencing and corrugated roof sheets.

Non-ferrous metals, such as aluminium and copper, are corrosion-resistant and are often chosen for this quality. Copper is used for electric wire, tubing for water supply and for flashing. Aluminium is most commonly used for corrugated roofing sheets, gutters and the accompanying nails. Using nails of the same material avoids the problem of corrosion caused by electrolytic action. Brass is a corrosion-resistant alloy of copper and zinc used extensively for building hardware.

Corrosion

Air and moisture accelerate corrosion in ferrous materials unless they are protected. Acids tend to corrode copper, while alkalis, such as that found in animal waste, portland cement and lime, as well as in some soils, cause rapid corrosion of aluminium and zinc. Electrolytic action, caused by slight voltages set up when dissimilar metals are in contact with each other in the presence of water, also encourages corrosion in some metals. Aluminium is particularly prone to electrolytic corrosion.

Corrosion can be reduced by the careful selection of metal products for the application, by reducing the time that the metal will be wet by preventing condensation and promoting good drainage, by avoiding contact between dissimilar metals, and by using corrosion-inhibiting coatings.

Corrosion-Inhibiting Coatings

Copper, aluminium, stainless steel and cast iron tend to form oxide coatings that provide a considerable amount of self-protection from corrosion. However, most other steels require protective coatings if they are exposed to moisture and air. Methods used include zinc coating (galvanizing), vitreous enamel glazing and painting. Painting is the only practical method for field application, although grease and oil will provide temporary protection.

Before painting, the metal surface must be clean, dry and free from oil. Both bituminous and oil-based paints with metallic oxide pigments offer good protection, if they are carefully applied in continuous layers. Two to three coats provides the best protection.

Building Hardware

Nails

A nail relies on the grip around its shank and the shear strength of its cross-section to give strength to a joint. It is important to select the right type and size of nail for any particular situation. Nails are specified by their type, length and gauge (the higher the gauge number, the smaller the shank diameter). Most nails are made from mild steel wire. In a corrosive environment, galvanized, copper-plated, copper or aluminium nails are used. A large number of nail types and sizes are available on the market. Below is a description of the nails most commonly used in farm buildings.

Round plain-headed nails or round wire nails are used for general carpentry work. As they have a tendency to split thin members, the following rule is often used: the diameter of the nail should not exceed 1 /7 of the thickness of the timber.

Table: Dimensions and approximate number per kilogram of commonly used sizes of round wire nails.

Length			
(inches)	(mm)	Diameter (mm)	Approximate number/kg
6	150	6.0	29
5	125	5.6	42
4	100	4.5	77
3	75	3.75	154
2.5	65	3.35	230
2	50	2.65	440
1.5	40	2.0	970
1	25	1.8	1 720

Lost-head nails have a smaller head, which can be set below the surface of the wood. Their holding power is lower because the head can be pulled through the wood more easily.

Panel pins are fine wire nails with small heads, used for fixing plywood and hardboard panels.

Clout or slate nails have large heads and are used for fixing tiles, slates and soft boards. Felt nails have even larger heads.

Concrete nails are made from harder steel, which allows them to be driven into concrete or masonry work.

Staples are U-shaped nails with two points, and are used mainly to fasten wires.

Roofing nails have a square, twisted shank and a washer attached to the head. Roofing felt or rubber may be used under the washer to prevent leakage. The nail and the washer should be galvanized to prevent corrosion. They are used for fixing corrugated-sheet materials and must be long enough to penetrate at least 20 mm into the wood. Alternatively, wire nails with used bottle caps for washers can be used.

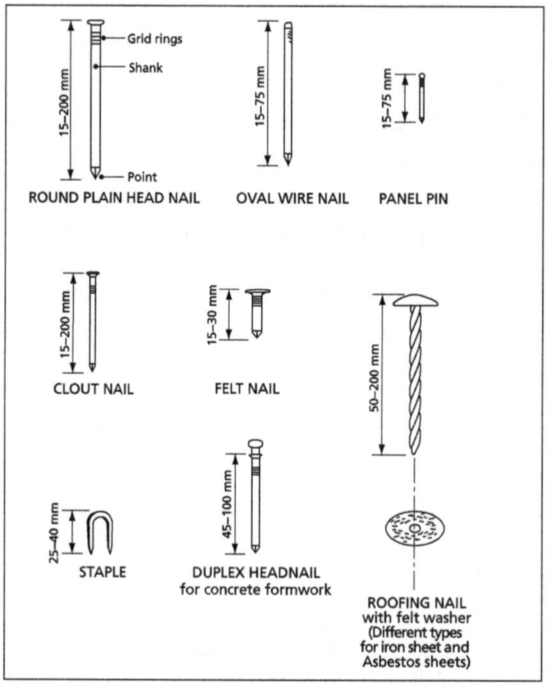

Types of nails.

Screws and Bolts

Wood screws have a thread, which gives them greater holding power and resistance to withdrawal than nails, and they can be removed easily without damaging the wood. For a screw to function properly, it must be inserted by rotation, and not by being driven with a hammer. It is usually necessary to drill a pilot hole for the shank of the screw. Screws made of mild steel are normally preferred because they are stronger. A wide range of finishes, such as galvanized, painted and plated, are available.

Screws are classified according to the shape of their head, as countersunk, raised, round or recessed (not slotted across the full width). Coach screws have a square head and are turned with a spanner. They are used for heavy construction work and should have a metal washer under the head to prevent damage to the wood surface. Screws are sold in boxes containing a gross (144 screws), and are specified by their material, finish, type, length and gauge. Unlike the wire gauge used for nails, the larger the screw-gauge number, the greater the diameter of the shank.

Bolts provide even stronger joints than either nails or screws. As the joint is secured by tightening the nut onto the bolt, in most cases the load becomes entirely a shear force. Bolts are used for heavy loads, such as at the joints in a gantry hoist frame, at the corners of a ring beam installed for earthquake protection, or to secure the hinges of heavy doors. Most bolts used with wood have a rounded head, with a square shank just under the head. Only one spanner is required for these 'coach' bolts. Square-head bolts, requiring two spanners, are also available. Washers help to prevent the nuts from sinking into the wood.

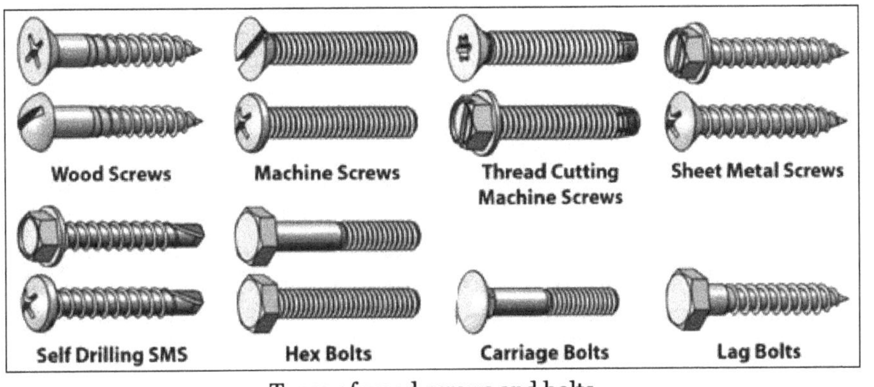

Types of wood screws and bolts.

Table: Conversion of screw gauge to millimetres.

screw gauge	0	1	2	3	4	5	6	7	8	9 10 12	14 16 18	20 24 28 32
Nominal diameter of unthreaded shank (mm)	1.52	1.78	2.08	2.39	2.74	3.10	3.45	3.81	4.17	4.52 4.88 5.59	6.30 7.01 7.72	8.43 9.86 11.28 12.70

Hinges

Hinges are classified by their function, length of nap and the material from which they are made, and come in many different types and sizes. Hinges for farm buildings are manufactured mainly from mild steel and are provided with a corrosion-inhibiting coating. The most common types are described below.

The steel butt hinge is commonly used for windows, shutters and small doors, as it is cheap and durable. If the pin can be removed from the outside, it is not burglarproof. The flaps are usually set in recesses in the door or window and the frame.

The H-hinge is similar to the butt hinge but is usually surface mounted.

The T-hinge is used mostly for hanging matchboarded doors. For security reasons, the strap of the T-hinge should be fixed to the door with at least one coach bolt, which cannot be unscrewed easily from the outside.

The band-and-hook hinge is a stronger type of T-hinge and is used for heavy doors and gates. This type is suitable for fabrication at the site or by the local blacksmith.

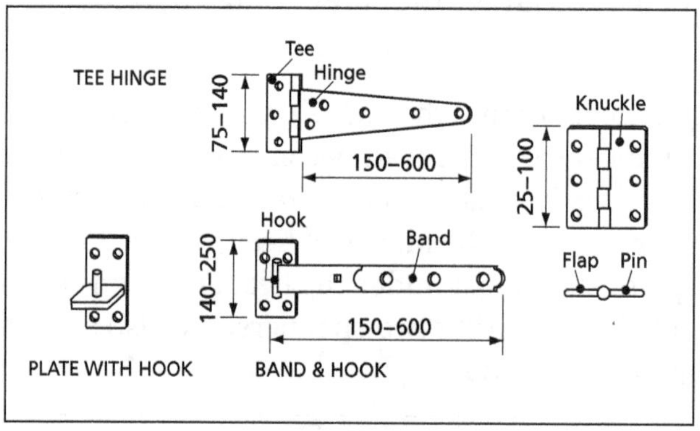

Types of hinges.

Locks and Latches

Any device used to keep a door in the closed position can be classified as a lock or latch. A lock is activated by means of a key, whereas a latch is operated by a lever or bar. Locks can be obtained with a latch bolt so that the door can be kept in a closed position without using the key. Locks in doors are usually fixed at a height of 1 050 mm. Some examples of common locks and latches used in farm buildings are illustrated in figure .

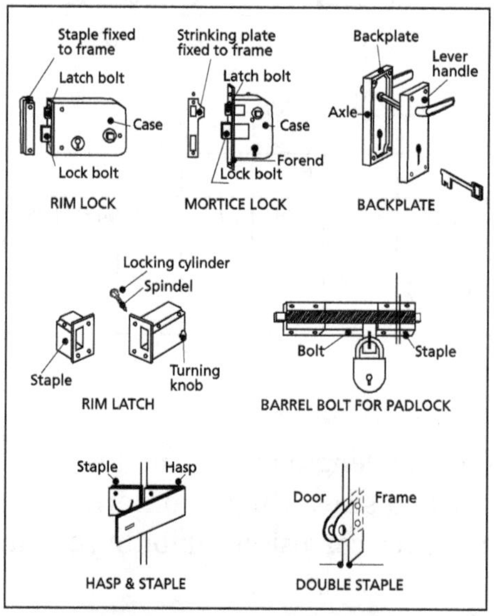

Types of locks and latches.

Glass

Glass suitable for general window glazing is made mainly from soda, lime and silica. The ingredients are heated in a furnace to about 1 500 °C, and fuse together in the molten state. Sheets are then formed by a process of drawing, floating or rolling. The ordinary glazing quality is manufactured by drawing in thicknesses ranging from 2 mm to 6 mm. It is transparent, with 90 percent light transmission. As the two surfaces are never perfectly flat or parallel, there is always some visual distortion. Plate glass is manufactured with ground and polished surfaces, and should be free of imperfections.

Glass in buildings is required to resist loads, including wind loads, impacts from people and animals, and sometimes thermal and other stresses. Generally the thickness increases with the area of the glass pane. Glass is elastic right up to its breaking point but, as it is also completely brittle, there is no permanent set or warning of impending failure. The support provided for glass will affect its strength performance. Glass should be cut to give a minimum clearance of 2 mm all around the frame to allow for thermal movements.

Plastics

Plastics are among the newest building materials, ranging from materials strong enough to replace metal, to foam-like products. Plastics are considered to be mainly organic materials derived from petroleum and, to a small extent, coal, which at some stage in their processing are plastic when heated.

The range of properties is so great that generalizations are difficult to make. However, plastics are usually light in weight and have a good strength-to-weight ratio, but rigidity is lower than that of virtually all other building materials, and creep is high.

Plastics have low thermal conductivity and thermal capacity, but thermal movement is high. They resist a wide range of chemicals and do not corrode, but they tend to become brittle with age.

Most plastics are combustible and may release poisonous gases in a fire. Some are highly flammable, while others are difficult to burn.

Plastics lend themselves to a wide range of manufacturing techniques, and products are available in many forms, both solid and cellular, from soft and flexible to rigid, and from transparent to opaque. Various textures and colours are available (many of which fade if used outdoors). Plastics are classified as:

- Thermoplastics, which always soften when heated and harden again on cooling, provided they are not overheated.
- Thermosetting plastics, which undergo an irreversible chemical change in which the molecular chains crosslink so that subsequently they cannot be appreciably softened by heat. Excessive heating causes charring.

Thermoplastics

Polythene is tough, waterproof and oilproof, and can be manufactured in many colours. In buildings, it is used for cold water pipes, plumbing and sanitary ware, and polythene film (translucent or black). Film should not be subjected unnecessarily to prolonged heat over 50 °C, or to direct sunlight. The translucent film will last for only 1–2 years if exposed to sunlight, but the carbon pigmentation of black film increases resistance to sunlight.

Polyvinyl chloride (PVC) will not burn and can be made in rigid or flexible form. It is used for rainwater goods, drains, pipes, ducts, electric cable insulation, etc.

Acrylics, a group of plastics containing polymethyl methacrylate, which transmit more light than glass and are easy to mould or curve into almost any shape.

Thermosetting Plastics

The main uses for thermosetting plastics in buildings are as impregnants for paper fabrics, binders for particle boards, adhesives, paints and clear finishes. Phenol formaldehyde (bakelite) is used for electrical insulating accessories, and urea formaldehyde is used for particle board manufacture.

In most cases, epoxy resins are provided in two parts: a resin and a curing agent. They are extremely tough and stable, and adhere well to most materials. Silicone resins are water-repellent and used for waterproofing in masonry. Note that fluid plastics can be very toxic.

Plastics used for Seepage Protection in Dams

Seepage from dams is a common problem. Occasionally, site conditions and the lack of local clay may require the use of synthetic liners, also called geomembranes, to line the dam and overcome the problem of seepage. If the correct product is selected and good installation procedures are used, only normal maintenance will be needed during the service life of the dam.

Three types of lining material are available: low- density plastic sheeting; woven polyethylene fabric; and high density polyethylene (HDPE) sheeting. All of these products are susceptible to degradation by introduced chemicals. The most suitable product for a specific job depends on a number of factors. These include: site conditions, cost, resistance to sunlight, strength, resistance to puncturing and the method of joining.

Low-Density Plastic Sheeting

Commonly referred to as 'builder's plastic', these sheets are normally black or orange in colour. Low-density plastic sheeting is manufactured in various thicknesses, with the recommended thickness being 0.2–0.3 mm. This sheeting punctures easily, as it is relatively thin. It is suitable for use on low slopes (less than 2:1, i.e. 2 metres horizontal to 1 metre vertical) and on sites free of sticks, small stones and abrasive materials. Extreme care needs to be taken during installation. Low-density plastic breaks down quickly in sunlight if it is not covered with soil.

Woven Polyethylene Fabric

This has a polyethylene coating on both sides, and is blue or green in colour. Woven polyethylene fabrics are generally the most suitable as dam liners in temperate countries, and are sold according to weight, not thickness. Fabric weight of about 250 grams per square metre is normally selected for farm dams. Heavier grades are required where puncturing is a concern. Although polyethylene fabrics are not UV resistant, their life expectancy can be increased to 15–20 years with soil cover.

It is normal practice to cover these fabrics with at least a 300 mm layer of soil. Owing to the need to provide this cover, sites with steep batters (greater than 2:1) are not suitable. For woven polyethylene fabrics used as dam liners, the batter slopes of the embankment and excavations should not exceed 2:1, although a gentler slope is preferable to ensure that any soil cover stays on the liner.

High-Density Polyethylene (HDPE) Sheeting

High-density polyethylene sheeting, which is black in colour, does not require soil cover, but it is the most expensive of lining materials. Installation is generally undertaken using fusion-weld joining equipment. Thicknesses range from 0.4 mm to 2.5 mm. It is suitable for sites where puncturing of cheaper products cannot be avoided, or where steep slopes (steeper than 2:1) preclude the use of other products.

HDPE is the most widely used geomembrane, and offers the most cost-effective liner for large, exposed, lining projects. This product has been used in landfills, wastewater treatment lagoons, animal waste lagoons, mining applications and water storage. It has the following advantages:

- Soil covering is not required;
- It has high overall chemical resistance and is resistant to ozone and uv;
- It is cost effective for large projects;
- It is suitable for potable water.

Plastic Components used with Dam Liners

In the process of lining dams, there may be needed one or more additional components that will ensure the longevity of the earth dam. These components include:

Geosynthetics

These are synthetic materials made from polymers (geomembranes are also classified under geosynthetics). When these materials are used together with dam liners (geomembranes), the service life of the earth dam is extended. The geosynthetics commonly used together with dam liners include:

- Geonets: These are open grid-like materials formed by a continuous extrusion of parallel sets of polymeric ribs intersecting at a constant acute angle. They are used in the design of drainage systems, particularly on slopes and are a viable alternative to the common sand and gravel systems.
- Geocells: These are constructed from polymeric strips, which are joined together to form a 3-dimensional network. They are used to stabilize the side slopes of dams and other earth structures. This usually involves filling the cells with soil.
- Geogrids: These are stiff or flexible polymer grid-like sheets with large uniformly distributed apertures. These apertures allow direct contact between soil particles on either side of the sheet. Their main use is to reinforce unstable soils.

Combination of two or more of the geosynthetics, e.g. geogrid and geomembrane, are referred to as geocomposites. Several combinations are possible and this area has attracted interest of many research establishments.

Degradable Erosion Mats

These are made of flexible erosion control blankets that are used to keep soil and seeds stable until vegetation completely covers the dam catchment. As they are made of organic materials, they eventually breakdown and become part of the soil.

Rubber

Rubbers are similar to thermosetting plastics. In the manufacturing process, a number of substances are mixed with latex, a natural polymer. Carbon black is added to increase strength in tension and to improve wearing properties.

After forming, the product is vulcanized by heating under pressure, usually with sulphur present. This process increases the rubber's strength and elasticity. Ebonite is a fully vulcanized, hard rubber.

Modified and synthetic rubbers (elastomers) are increasingly being used for building products. Unlike natural rubbers, they often have good resistance to oil and solvents. One such rubber, butyl, is extremely tough, has good weather resistance, excellent resistance to acids and very low permeability to air. Synthetic rubber fillers and nail washers are used with metal roofing.

Bituminous Products

These include bitumen (asphalt in the United States), coal tar and pitch. They are usually dark brown or black and, in general, they are durable materials that are resistant to many chemicals. They resist the passage of water and water vapour, especially if they have been applied hot.

Bitumen occurs naturally as rock asphalt or lake asphalt, or can be distilled from petroleum. It is used for road paving, paint, damp-proof membranes, joint filler, stabilizer in soil blocks, etc.

Paints

Paint preserves, protects and decorates surfaces, and enables them to be cleaned easily. All paints contain a binder that hardens. Other ingredients found in various paints include: pigments, strainers, extenders, driers, hardeners, thinners, solvents and gelling agents. Some water-thinned paints contain emulsifiers.

Owing to the cost involved, few buildings in rural areas are painted. When paint can be afforded, priority should be given to painting surfaces likely to rust, rot or decay because of exposure to rain or dampness, and to rooms such as a kitchen or a dairy, where hygiene demands easily cleaned surfaces. White and other light colours reflect more light than dark colours, and can be used in a sitting room or a workshop to make the room lighter.

Painting

Adequate preparation of the surface to be painted is essential. The surface should be smooth (not shiny, because this would not give good adhesion), clean, dry and stable. Old, loose paint should be brushed off before a new coat is applied. Most commercial paints are supplied with directions for use, which should be read carefully before the work is started. The paint film is usually built up in two or more coats.

Priming paints are used for the first coat, to seal and protect the surface and to give a smooth surface for subsequent coats. They are produced for application to wood, metal and plaster.

Undercoating paints are sometimes used to obscure the primer, as a further protective coating and to provide the correct surface for the finishing paint.

Finishing paints are produced in a wide range of colours and finishes (e.g. matt, semi-matt or gloss). Some commonly used types of paint for farm structures are detailed below, but many others are manufactured with special properties, making them water- and chemical-resistant, heat-resistant, fire-retardant, anti- condensation, fungicidal or insecticidal, for example.

Estimation of Quantities of Paint Required

The volume of paint required for a particular paint job can be determined from knowledge of the following:

1. Surface area of the surface(s) to be painted.
2. Spreading rate of the paint being used.
3. The number of coats needed.

Spreading Rates

The spreading rate of paint is the area that a specific volume of paint will cover at a specified film thickness. Two standard measurements are used to describe the film thickness of a coating: mils and microns. A micron is a metric system measurement equal to 0.001 millimetres.

The spreading rate in microns may be calculated as follows:

Any liquid will cover 1 000 square metres per litre at 1 micron wet. Therefore, a 100 percent volume/solids material will cover 1 000 m2 per litre dry when applied at 1 micron wet and, because it is 100 percent solids, it will yield a 1 micron-thick dry film. However, if a coating is less than 100 percent solids, then the dry film thickness will be thinner because the volatile portion of the volume will evaporate and leave the film, thereby reducing the dried film volume or thickness.

Assuming a material has 42 percent volume of solids, the area that it will cover when a dry film thickness of 38.1 microns is required may be calculated as follows:

1 000 m^2 / litre × 0.42 = 420 m^2 / litre at 1 micron thickness.

Assuming 38.1 microns dry is desired, then:

420 m^2 / litre / 38.1 = 11.0 m^2 / litre.

A coating with 42 percent volume solids, applied at 11.0 m^2 / litre, will produce a dry film that is 38.1 microns thick.

Example:

The living room walls require painting, excluding the ceiling. The walls are 3 metres high, with a total of 18 metres of wall length. The total door and window area is 3 m2. If a spreading rate of 11 square metres per litre is used, and only two coats of paint are required, work out how much paint is needed.

To work out how much paint is needed:

- Take the surface area = $(3\times18) - 3 = 51\ m^2$
- The spreading rate is $11\ m^2$ / litre
- The number of coats needed = 2
- The required litres of paint = ($51m^2$ / $11m^2$ / litres) × 2 coats = 9 litres in total (i.e. a 10-litre pail).

Oil and Resin-Based Paints

Oil paints are based on naturally drying oils (e.g. linseed oil). They are being gradually replaced by alkyd and emulsion paints.

Alkyd paints are oil-based paints, modified by the addition of synthetic resins to improve durability, flexibility, drying and gloss. They are quite expensive.

Synthetic resin paints contain substantial proportions of thermosetting resins, such as acrylics, polyurethane or epoxides, and are often packed in two parts. They have excellent strength, adhesion and durability, but are very expensive.

Bituminous paints are used to protect steelwork and iron sheeting from rust, and to protect wood from decay. They are black or dark in colour, and tend to crack in hot sunlight. They can be over-painted with ordinary paint only after a suitable sealer has been applied.

Varnishes are either oil/resin or spirit-based and used mainly to protect wood with a transparent finish, but protection is inferior to opaque finishes. Spirit- based varnish is used only for interior surfaces.

Water-Based Paints

Non-washable distemper consists of chalk powder, mixed with animal glue dissolved in hot water. It is cheap, but easily rubbed or washed off, and therefore suitable only for whitening ceilings.

Washable distemper (water paint) consists of drying oil or casein, emulsified in water with the addition of pigments and extenders. Hardening is slow but, after a month, it can withstand moderate scrubbing. It weathers fairly well outdoors and is reasonably cheap.

Whitewash (limewash) consists of lime mixed with water. It can be used on all types of wall, including earth walls, and is cheap, but its lack of water resistance and poor weathering properties make it inferior to emulsion paint for outdoor surfaces. However, the addition of tallow or cement gives some degree of durability for external use. Whitewash can be made in the following way:

- Mix 8 litres (9 kg) of quicklime with about 18 litres of boiling water, adding the water slowly and stirring constantly until a thin paste results.

- Add 2 litres of salt and stir thoroughly.
- Add water to bring the whitewash to a suitable consistency.
- If external quality is required, add a handful of cement per 10 litres of whitewash just before use.

In emulsion paints, the pigments and binder (vinyl, acrylic, urethane or styrene polymers) are dispersed as small globules in water. They harden quickly, are quite tough and weather-resistant, and the cost is moderate. Although they adhere well to most supports, because they are permeable an oil-based primer may be required to seal porous exterior surfaces.

Cement-based paints are often used for exteriors, and are quite inexpensive. They contain white portland cement, pigments (if other colours are desired) and water-repellents, and are sold in powder form. Water is added just before use to obtain a suitable consistency. Paint that has thickened must not be thinned further. It adheres well to brickwork, concrete and renderings, but not to timber, metal or other types of paint. Surfaces should be dampened before painting.

Cement slurries make economical surface coatings on masonry and concrete, but earth walls that shrink and swell will cause the coating to peel off. Slurries are mixtures of cement and/or lime, clean fine sand and enough water to make a thick liquid. A good slurry can be made using 1 part cement, 1 part lime and up to 4 parts sand. It is applied on the dampened surface with a large brush or a used bag, hence the name 'bag washing'.

Properties of Building Materials

For a material to be considered as building material, it should have required engineering properties suitable for construction works. These properties of building materials are responsible for its quality and capacity and helps to decide applications of these material.

Such properties of building materials are categorized as follows:

- Physical properties
- Mechanical properties
- Chemical properties
- Electrical properties
- Magnetic properties
- Thermal properties

Physical Properties of Building Materials

These are the properties required to estimate the quality and condition of the material without any external force. The physical properties of engineering materials are as follows.

- Bulk density

- Porosity
- Durability
- Density
- Density index
- Specific gravity
- Fire resistance
- Frost resistance
- Weathering resistance
- Spalling resistance
- Water absorption
- Water permeability
- Hygroscopicity
- Coefficient of softening
- Refractoriness.

Bulk Density of Building Materials

Bulk density is the ratio of mass to the volume of the material in its natural state that is including voids and pores. It is expressed in kg/m^3. Bulk density influences the mechanical properties of materials like strength, heat and conductivity etc. bulk density values of some of the engineering materials are given below.

Building material	Bulk density (kg/m^3)
Brick	1600 – 1800
Sand	1450 – 1650
Steel	7850
Heavy concrete Light concrete	1800 – 2500 500 – 1800
Granite	2500 – 2700

Porosity of Building Materials

Porosity gives the volume of the material occupied by pores. It is the ratio of volume of pores to the volume of material. Porosity influences many properties like thermal conductivity, strength, bulk density, durability etc.

Durability of Building Materials

The property of a material to withstand against the combined action of atmospheric and other factors is known as durability of material. If the material is more durable, it will be useful for longer life. Maintenance cost of material is dependent of durability.

Density of Building Materials

Density is the ratio of mass of the material to its volume in homogeneous state. Almost all the physical properties of materials are influenced by its density values. Density values of some building materials are given below.

Material	Density (kg/m^3)
Steel	7800 – 7900
Brick	2500 -2800
Granite	2600 – 2900

Density Index

Ratio of bulk density of material to its density is termed as density index. Hence it gives the volume of solid matter in the material. In nature, fully dense material is not available so, density index is always less than 1 for any building material.

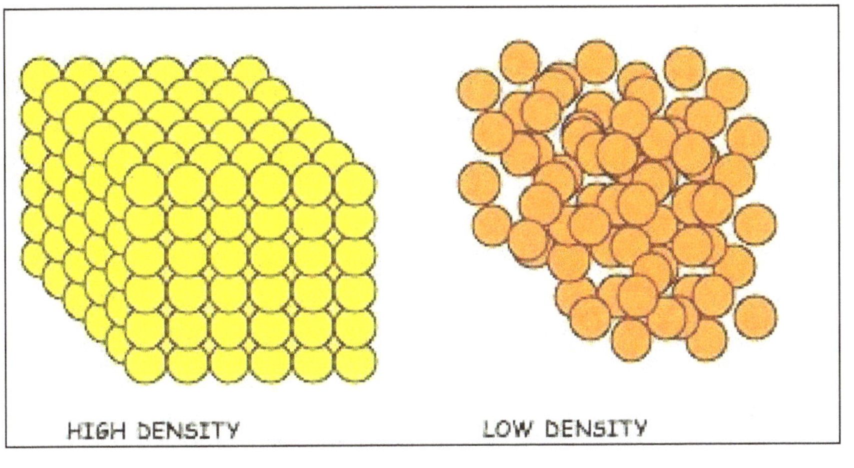

Specific Gravity of Building Materials

Specific gravity is the ratio of mass of given substance to the mass of water at 4°C for the equal volumes. Specific gravity of some materials is listed below.

Material	Specific gravity
Steel	7.82
Cast iron	7.20
Aluminum	2.72

Fire Resistance of Building Materials

The ability to withstand against fire without changing its shape and other properties. Fire resistance of a material is tested by the combined actions of water and fire. Fireproof materials should provide more safety in case of fire.

Frost Resistance

The ability of a material to resist freezing or thawing is called frost resistance. It is depends upon the density and bulk density of material. Denser materials will have more frost resistance. Moist material have low frost resistance and they lose their strength in freezing and become brittle.

Weathering Resistance

The property of a material to withstand against all atmospheric actions without losing its strength and shape. Weathering effects the durability of material. For example corrosion occurs in iron due to weathering. To resist this paint layer is provided.

Spalling Resistance

The ability of a material to undergo certain number of cycles of sharp temperature variations without failing is known as spalling resistance. It is the dependent of coefficient of linear expansion.

Water Absorption

The capacity of a material to absorb and retain water in it is known as water absorption. It is expressed in % of weight of dry material. It depends up on the size, shape and number of pores of material.

Water Permeability

The ability of a material to permit water through it is called water permeability. Dense materials like glass metals etc. are called impervious materials which cannot allow water through it.

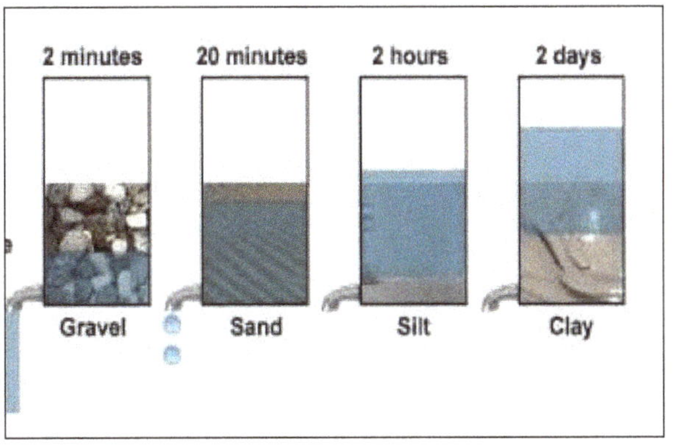

Hygroscopicity

Hygroscopicity is the property of a material to absorb water vapor from the air. It depends on the relative humidity, porosity, air temperature etc.

Coefficient of Softening

Coefficient of softening of a material is the ration of compressive strength of a saturated material to its compressive strength in dry state. It affects the strength of water absorbent materials like soil.

Refractoriness

The property of a material which cannot melts or lose its shape at prolonged high temperatures (1580 °C or more).

Example: Fire clay is high refractory material.

Mechanical Properties of Building Materials

Mechanical properties of the materials are find out by applying external forces on them. These are very important properties which are responsible for behavior of a material in its job. The mechanical properties are,

- Strength
- Hardness

- Elasticity
- Plasticity
- Brittleness
- Fatigue
- Impact strength
- Abrasion resistance
- Creep.

Strength of Building Materials

The capacity of a material to resist failure caused by loads acting on it is called as strength. The load may be compressive, tensile or bending. It is determined by dividing the ultimate load taken by the material with its cross sectional area. Strength is an important property for any construction materials. So, to provide maximum safety in strength, factor of safety is provided for materials and it is selected depending on nature of work, quality of material, economic conditions etc.

Hardness of Building Materials

The property of a materials to resist scratching by a herder body. MOHS scale is used to determine the hardness of a materials. Hardness is most important to decide the usage of particular aggregate. It also influences the workability.

Elasticity of Building Materials

The capacity of a material to regain its initial shape and size after removal of load is known as elasticity and the material is called as elastic material. Ideally elastic materials obey Hooke's law in which stress is directly proportional to strain. Which gives modulus of elasticity as the ratio of unit stress to unit deformation. Higher the value of modulus of elasticity lower the deformations.

Plasticity

When the load is applied on the material, if it will undergo permanent deformation without cracking and retain this shape after the removal of load then it is said to be plastic material and this property is called as plasticity. They give resistance against bending, impact etc.

Examples: steel, hot bitumen etc.

Brittleness

When the material is subjected to load, if it fails suddenly without causing any deformation then it is called brittle material and this property is called as brittleness.

Examples: concrete, cast-iron etc.

Fatigue

If a material is subjected to repeated loads, then the failure occurs at some point which is lower than the failure point caused by steady loads. This behavior is known as fatigue.

Impact Strength

If a material is subjected to sudden loads and it will undergo some deformation without causing rupture is known as its impact strength. It designates the toughness of material.

Abrasion Resistance

The loss of material due to rubbing of particles while working is called abrasion. The abrasion resistance for a material makes it durable and provided long life.

Creep

Creep the deformation caused by constant loads for long periods. It is time dependent and occurs at very slow rate. It is almost negligible in normal conditions. But at high temperature conditions creep occur rapidly.

Chemical Properties of Building Materials

The properties of materials against the chemical actions or chemical combinations are termed as chemical properties. And they are:

- Chemical resistance.
- Corrosion resistance.

Chemical Resistance of Building Materials

The ability of a construction materials to resist the effects by chemicals like acids, salts and alkalis is known as chemical resistance. Underground installations, constructions near sea etc. should be built with great chemical resistance.

Corrosion Resistance

Formation of rust (iron oxide) in metals, when they are subjected to atmosphere is called as corrosion. So, the metals should be corrosive resistant. To increase the corrosion resistance proper measures should be considered. Otherwise it will damage the whole structure.

Electrical Properties of Building Materials

The properties of a material to conduct or to resist electricity through them are electrical properties of material. For example, wood have great electric resistance and stainless steel is a good conductor of electricity.

Magnetic Properties of Building Materials

The magnetic properties of materials like permeability, hysteresis etc. are required in the case of generators etc. iron is magnetic material and aluminum is non-magnetic material.

Thermal Properties of Building Materials

- Thermal capacity
- Thermal conductivity
- Thermal resistivity
- Specific heat

Thermal Capacity of Building Materials

Thermal capacity is the property of a material to absorb heat and it is required to design proper ventilation. It influences the thermal stability of walls. It is expressed in J/N °C and it is calculated by below formula.

Thermal capacity, $T = [H/(M(T_2 - T_1))]$

Where H = quantity of heat required to increase the temperature from T_1 to T_2

T_1 = Initial temperature

T_2 = Final temperature

M = Mass of material in N.

Thermal Conductivity

The amount of heat transferred through unit area of specimen with unit thickness in unit time is termed as thermal conductivity. It is measured in kelvins. It depends on material structure, porosity, density and moisture content. High porous materials, moist materials have more thermal conductivity.

Thermal Resistivity

It is the ability to resist heat conduction. And it is the reciprocal of thermal conductivity. When it is multiplied by thickness of material it gives thermal resistance. Thermal resistivity of soil varies from 30 to 500 °C-cm/W.

Specific Heat

Specific heat is the quantity of heat required to heat 1 N of material by 1°C. Specific heat is useful when we use the material in high temperature areas. Specific heat values of some engineering materials are given below.

Material	Specific heat J/N oC
Steel	0.046 x 103
Wood	0.239 to 0.27 x 103
Stone	0.075 to 0.09 X 103

3

Brick Masonry Construction

Brick refers to the building material which is used to make pavements, walls and several other elements in masonry construction. The form of construction where bricks are placed together in a systematic manner and mortar is added to the gaps is known as brick masonry. This chapter discusses brick masonry in detail as well as the different types of bricks.

Brick

Bricks are one of the oldest and most popular building materials which always used in building Construction. Bricks are blocks of tampered clay moulded to suitable shapes and sizes while it is still in plastic condition, dried in the sun and burnt, if desired so as to make them more strong, hard and durable.

Types of Bricks

There are two main types of Bricks.

Unburnt Bricks

These bricks are dried through sun light. It is also called Sun-dried Bricks. They have very low strength as compared to other types of bricks.

It is believed that first-time bricks were used, in sun-dried form (unburnt Bricks), in Egypt some 6000 years ago.

Burnt Bricks

The bricks which are used commonly in today's age are burnt bricks. They are prepared and burnt in a kiln.

They have high strength as compared to unburnt bricks. They are further classified into the following categories.

First Class Bricks

These are 19 x 9 x 9 cm in size. They are made from good earth, free from saline deposits. They should be thoroughly burnt.

They should be of good color. They should be of regular shape with square edges and parallel faces.

These bricks are free from flaws, cracks, chips, stones, etc. They should give a ringing sound when two bricks are struck together.

Its compressive strength shall not be less than 140 kg/cm². And they shall not absorb more than 20% of water when immersed in water for 24 hours.

Use: Excellent for all types of construction in the exterior walls. They are also suitable for flooring.

Second Class Bricks

Second class bricks are also fully burnt and give a clear ringing sound when struck together. Slightly irregularities in shape, size or color are accepted.

Its compressive strength shall not be less than 70 kg/cm², and absorption value should not be greater than 22 percent when soaked for 24 hours in water.

Slight difference in the structure on fractured surfaces is admissible.

Use: For exterior work when plastering is to be done. And can also be used for interior works but they may not be used for flooring.

Third Class Bricks

These are not burnt so fully as in previous two cases but are generally of uniform reddish yellow color.

Defects in uniformity or shape are tolerated. On striking together, they produce a dull thud sound.

Its compressive strength lies between 35 – 70 kg/cm² and absorption between 22 – 25 percent.

Use: They are used mostly in the ordinary type of construction and in dry situations.

Fourth Class Bricks

These types of bricks are irregular in shape and dark in color which is due to overburning.

They are quite strong in compressive strength, generally above 150 kg/cm² and low in porosity and absorption.

Use: Despite their high strength, these types of bricks are unfit for use in building construction. This is because of their distorted shape and irregular size.

They are, however, very commonly used in a

broken form, in road construction, foundations and floors as a coarse aggregate material.

Special Types of Bricks

Ordinary building bricks are typically rectangular in shape, solid in structure and made from a Suitable type of brick-clays.

In building construction, however, bricks of modified Shapes, porous or hollow structure and also made from materials other than clay are also sometimes used.

A familiarity with this Special classification of bricks is quite important for a construction engineer and an architect.

Classification of Bricks based on Modification in Shape

In construction, there are positions where a perfect rectangular shape brick will not be suitable. It has to be modified.

Such bricks are required in the plinth, the corners of the walls and at the copings.

There is a long list of such Special-Shaped bricks.

Squint Brick

They are cut on one corner at an angle of other than 90 degrees. They are required for giving shape to an exterior or interior corner in a wall.

Splay or Can Brick

These have a level or portion taken off, width wise, length wise, or in both directions. The various shapes grouped under splay bricks are made for use in jambs of doors and windows and also in plinths.

Coping Bricks

They are used for coping on walls in order to give them a nice appearance and also for easy drainage of water.

When a coping is to be provided to a wall, a special shape may be desired. The chamfered, the half round, and the saddleback bricks are some common coping bricks.

Bullnose

Bullnose bricks are used at turns of the wall so that round corners are obtained. It is a standard brick having one edge rounded.

Cownose

It is similar to bullnose, but it has both edges rounded on one side. It may also be called double bullnose bricks. They have the same use as bullnose, but they just give additional roundness.

Classification of Bricks based on Cavities

Perforated Bricks

These form a modem class of building bricks that have many advantages over the ordinary solid brick.

In perforated bricks, cylindrical, rounded or rectangular holes are made in the bricks after the molding stage. These holes are called perforations.

They are properly spaced from the side of the brick, and the minimum distance is 15 mm. Further, a distance between any two perforations is not less than 10 mm.

The volume of the perforations may be as much as 20 – 50 percent of the total volume of the brick.

They may be larger in size than ordinary building brick which is a distinct advantage in that work output of a Mason will increase considerably with the use of perforated bricks.

The major advantages that perforated bricks offer over ordinary bricks are:

1. They are light in weight.
2. Less quantity of clay is required for their manufacture.
3. Less time is required for drying and burning of perforated bricks.

4. These offer better resistance against rain penetration and better insulation against heat. As such they are ideally suited for tropical countries.

5. Efflorescence is least in perforated bricks.

6. These are especially suitable for the construction of brick panels in multi-story structures. Perforated bricks have as yet to find popularity in all countries although they are already widely used in Germany, France, and America.

Already there is a trend for their use.

The manufacture of perforated bricks requires slightly sophisticated technology, and that is the main reason for lesser use in all over the world.

Hollow Brick

Hollow bricks also called cavity bricks or cellular bricks they have well-defined sets of cavities with specified dimensions made in the body of the brick.

As a result, their net weight may be only one-third to one-half of the solid portion. It is important that the thickness of the brick wall near the cavity should not be less than 2 cm.

The hollow bricks are made from a special type of brick clay (which should have higher clay content).

They offer the following advantages over the ordinary types of bricks:

1. Being light in weight, they can be handled more conveniently, and the output of the mason may be three to four times compared to ordinary bricks.

2. They offer better insulation against heat and sound.

3. They are ideal and economical for non-load-bearing walls, e.g., in partition walls.

Channel or Gutter Bricks

They are actually special purpose bricks baring a continuous central semi-circular cavity or depression running through their length.

Very often they are glazed to make them impervious: They are mostly used in the laying of drains.

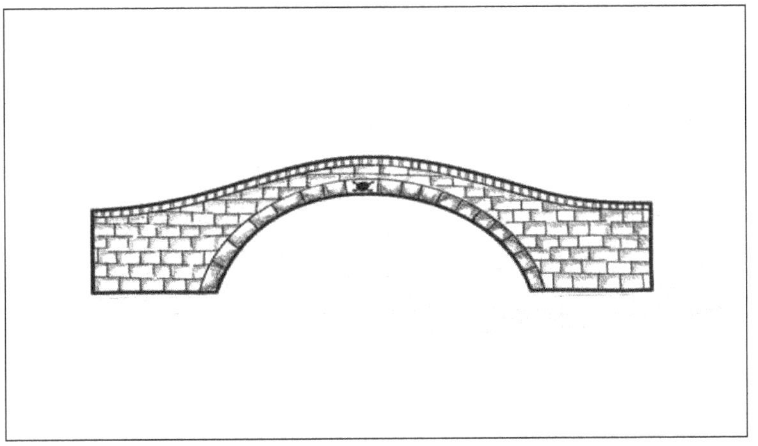

Classification of Bricks based on Composition

Sand Lime Bricks

These are building bricks made from sand and lime as the raw materials instead of clay. The clay content may be only negligible.

They differ from ordinary clay bricks not only in composition but also in the method of manufacture. They are, however, similar in shape and size to the ordinary types of bricks.

Manufacture: The principal raw materials for sand-lime bricks are:

1. Sand: It must be free from harmful impurities like chloride, iron oxides, black minerals and organic matter. Sand forms around 90 percent of sand-lime bricks.
2. Lime: It is used in the form of a slaked (hydrated) lime. Its content varies from 8-12 percent.

The slaked lime used for brick making has high calcium-content and should be free from magnesium.

Besides the two essential components, some clay (up to four percent) in finely divided form is desirable. Pure, salt-free water (not sea water) is another requirement.

For giving a required color to the sand-lime bricks, some pigments are added in small percentages to the ingredients at the mixing stage.

Among these pigments, the following are used commonly:

Iron-oxides for red and brown color.

Chromium Oxide for green color.

Ochre for yellow color.

Carbon black for Grey and black color.

Molding: After mixing the finely powdered raw materials in the desired proportion, a damp mixture is obtained adding 2-3 per cent clean water.

From this damp mixture, brick-shaped units are molded using rotary press. Pressures applied range from 300-600 kg/cm^2.

Such high-pressure results in highly compressed and dense sand-lime bricks which are almost dry at this stage.

Autoclaving: The molded units are put into an autoclave (a steel cylinder with the closed end where heating is done by steam under pressure).

In the autoclave, the bricks are treated for 6-12 hours under steam pressures between 8-16 kg/cm^2.

It is in the autoclave that sand and lime react chemically and form a chemical compound, calcium-silicate. That is the chemical composition of the sand-lime brick.

The sand-lime bricks taken out from the autoclave are ready for use.

Properties of Sand Lime Bricks

1. The sand-lime bricks have a very smooth and uniform finish and a pleasing appearance.
2. They are quite dense, strong and hard.
3. They are least porous and hence free from efflorescence.

Because of these properties, sand-lime bricks offer many advantages.

1. They are uniform in shape, size and finish and hence require no plastering.
2. The quantity of plaster when required is quite less because of the smooth surface of the brick.
3. They are water repelling and hence free from absorption related defects of ordinary bricks.
4. Since the materials required for sand-lime bricks are also quite common in occurrence, these can provide a suitable alternative to clay-bricks.

We should know that most clay used in making bricks is from agricultural lands and hence its conservation will help in increasing food production.

Among the major disadvantages of sand-lime bricks:

1. They can be manufactured only by using mechanized methods. Hence their production at village levels is almost impossible at present.
2. They are unsuitable for foundations and paving uses. In the first case, they get damaged in the presence of water, and in the second case, it is because they have poor resistance to abrasion.

The Scope of Use: In Western countries, sand-lime bricks are used quite widely. Sooner or later, they will have to be introduced in other countries too on a large scale.

Fire or Refractory Bricks

This is a separate group of bricks which is capable of withstanding in very high temperatures without undergoing any deformation in shape or size and without reacting with the material of a particular composition.

Their use is restricted to the making of inner walls of furnaces for the manufacture of metals and for similar high-temperature applications.

Classification

The firebricks are classified on the basis of their reactivity towards melts at high temperatures:

a. *Acidic Bricks:*They are resistant to the melts of acidic composition (but will react with the melts of basic composition).

   Example: Fire-clay bricks, silica bricks.

b. *Basic Bricks:*They are resistant to the melts of basic composition. Hence they will not be suitable for use in those furnaces where acidic melts are being heated.

   Examples: Magnesia bricks, Bauxite bricks.

c. *Neutral Bricks:* They are non-reactive to both the acidic and basic melts. Hence these find applications in heating either type of melts.

   Examples: Chromite bricks, Chrome-magnesite bricks.

Manufacture of Fire Bricks

They are manufactured in the same manner as a building Brick. The four stages are involved in their manufacture process are:

- Moulding
- Drying
- Burning
- Cooling

The Raw Materials are first crushed in to the required size and then blended are mixed thoroughly in the presence of required quantity of water.

After this process brick are molded manually or by using machines such as "presser".

The molded bricks are dried and then burnt in a very high temperatures, generally between 1600 to 2000 centigrade.

After this process they are allowed to cool very gradually.

Fire Clay Bricks

These constitute an important class of refractory materials of acidic group. Raw material for its manufacturing can withstand at very high temperatures without fusing or softening.

Such clay are often available under the coal layers in nature.

Silica (65-75 percent) and alumina (25-35 percent) are the two main constituents of fire clay.

They are free from impurities like oxides of *calcium, magnesium* and *iron*. The maximum permissible upper limit for all such impurities is 5 percent.

These are manufactured in a manner similar to the common building brick. The selected clay is crushed to the fine powder and molded into the brick unit.

These are than dried and burnt in continuous kiln at very high temperature (1600 to 1900 centigrade).

They are cooled gradually before taking out from the kiln.

Types

They are divided into three types on the basis of temperature.

- High Duty. (1482 to 1648 Centigrade).
- Medium Duty (1315 to 1481 Centigrade).
- Low Duty (870 to 1314 Centigrade).

Properties

They have high resistance to palling, high bearing capacity and low coefficient of thermal expansion.

Uses

They are commonly used as a lining material in steel making furnaces and reverberatory furnaces.

Silica Fire Bricks

These are made up mostly of silica, which may range from 95 percent or above. The remaining material is generally calcium oxide that act as binding agent.

These are acidic in character. Their raw materials is pure quartz (SiO_2), or sandstone of high silica content.

The molded unit are burnt in about 1500 Centigrade. They are extensively used in bessemer converter as lining material because slag in that case are acidic in nature.

Magnesite Fire Bricks

They are classed among the basic refractories. There raw material is magnesium oxide. The major source for it is dolomite rock.

Bauxite Fire Brick

They also form a very important class of Basic Refractories. They are made from the rock bauxite ($Al_2O_32H_2$) which are mixed with some clay (Fire Clay type).

Chromite Fire Bricks

They are belong to the neutral class of refractories. Raw material used for these Fire Bricks is a mineral called chromite.

The mineral is double oxide of chromium and iron in its composition. These types Refractory bricks are capable of resisting both the acidic and basic environments in a furnace.

These type of fire brick are commonly used in the steel making furnaces.

Brick Masonry

Brick masonry is a highly durable form of construction. It is built by placing bricks in mortar in a systematic manner to construct solid mass that withstand exerted loads. There are several types of bricks and number of mortars which can be used to construct brick masonry.

The bond in brick masonry, which adheres bricks together, is produced by filling joints between bricks with suitable mortar. Special cautions shall be practiced while mortar is mixed and placed since it greatly affect the performance and durability of masonry structure.

Types of Brick Masonry Work

Brick Work in Mud

- The mud is used to fill up various joints brick masonry work.
- Thickness of the mortar joint is 12 mm.
- It is the cheapest type of brick masonry
- Employed for construction of walls with maximum height of 4 m.

Brick work in mud.

Brick Work in Cement

This type of brick masonry is construction by laying bricks in cement mortar rather than mud which is used in brick work in mud. There are three major classes of brick work in cement which are summarized in Table.

Table: Different classes of brick work in cement and their descriptions.

Classes	Descriptions
First Class	Cement of lime mortar is used, The surface and edges of bricks are sharp, And the thickness of mortar joints doesn't exceed 10mm

Second Class	Ground moulded bricks are used, Bricks are rough and shape is slightly irregular, The thickness of mortar joint is 12 mm.
Third Class	Bricks are not hard ,rough surface with distorted shape, Used for temporary structures, Used in places where rainfall is not heavy.

Brick work in cement.

Brick Masonry Construction

Materials and Equipment used in Brick Masonry Construction

1. Mortar Mix or Mason Mix Bricks
2. Tape measure
3. Hammer
4. Hose, level, or theodolite
5. Trowel Level
6. Wheelbarrow
7. Goggles
8. Jointer
9. And other equipment according to project ans personal preferences.

Brick Masonry Construction Preparations

1. Check the level of the ground using level, theodolite or transparent hose level.
2. Set the layout of the structure.

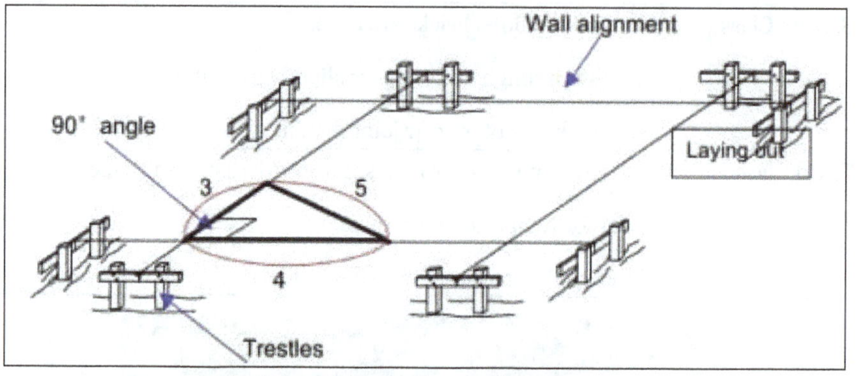

Setting out layout of masonry structure.

1. Trace of building axis and wall alignment using gypsum powder, chalk, or similar, marking the trenches for foundation.
2. After that,install foundation wall, cure foundation for minimum two days before beginning of brick masonry construction.
3. Distribute bricks in several stacks along project site to cut time and effort later.
4. Wet bricks few hours prior to the work. Not only does this avoid absorbing too much water from mortar but also improve adherence of bricks and mortar.

Brick Masonry Construction Procedure

1. Initially, mix the mortar with water and blend it until a smooth and plastic mortar is produced.

Mortar Preparation.

2. After that, place the mortar on foundation line evenly using trowel (25mm thickness and one brick wide is recommended for laid mortar).
3. Then, lay the first course of stretcher bricks in the mortar. Start with second brick, apply mortar to the head joint end of each brick, After that shove the bricks into place firmly so that the mortar is squeezed out of all side of the joints.

Laying bricks.

4. Utilize a level to examine the course for correct height. ensure that bricks are plumb and level.

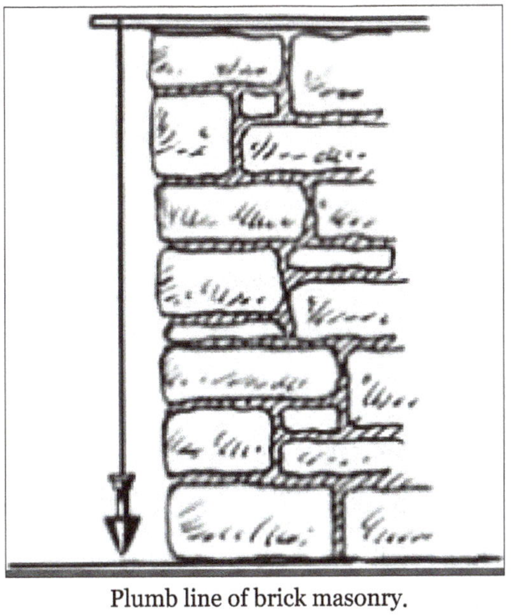

Plumb line of brick masonry.

Checking level of brick masonry.

5. Place another mortar line alongside the first course, then begin laying the second course.
6. Use the two half bricks to begin the second to ensure that the first two courses are staggered for structural purposes.
7. To finish the second course of the lead, lay three header bricks and make sure that they are plumb and level.
8. The third and fifth courses consists of stretchers similar to the first course. The fourth course begins with single header, followed by stretchers. Use the level to make sure that the lead is true on each course. Lastly, this pattern of brick laying is used till the target height is reached.

Points Considered in Supervising Brick Masonry Constructions

Brick bat.

The following points should be observed in the construction of brick masonry:

1. Use good quality bricks.
2. Ensure that brick courses are perfectly horizontal.
3. Verticality of the wall should be ensured by frequently checking with plumb-bob.
4. Whenever work is stopped brick masonry should be left with toothed end.
5. Use of brick bats should be avoided.
6. Raising walls by more than 1.5 m in one day shall be prevented.
7. Raise face joints to a depth of 12 to 20mm so as to be used as a key for plastering or pointing.
8. Brick masonry should be regularly cured for 2 weeks.
9. The thickness of mortar joints shall be 10 mm both horizontally and vertically.

4

Stone and Concrete Construction

The construction which makes use of stones that are bonded together by mortar is known as stone masonry. The structural material consisting of a hard and chemically inert particulate substance which is bonded together by cement and water is called concrete. The diverse applications of stones and concrete in construction have been thoroughly discussed in this chapter.

Stone

Stone is a great floor finish and is affordable in many countries; There is a great advantage to stone that is almost unique to floor finishes: you can polish it, and thus make it look and feel like new, at any time in its life.

Granite is a volcanic rock (it was originally lava, that cooled to form solid rock) that has the following properties:

- It is very hard, strong, and abrasion resistant.
- It is resistant to acids.
- It can be polished to a mirror-like smoothness.

These properties make it a great choice as a floor or countertop finish. It can also be used to clad walls. However it is available mainly in dark colours: black, red, grey. This darkness in colour does tend to limit its use in certain areas.

Its surface can also be worked to produce a variety of textures other than smooth: granite can be flamed, water blasted, sand blasted, bush hammered, or tumbled. These rough finishes are mostly used outdoors, on pathways. The famous cobblestones of Europe are granite, for instance.

Marble is a metamorphic rock (meaning that it was made by the intense pressures and heat deep within the earth), and has the following properties:

- Most marbles are soft, and not very abrasion resistant.
- They are not resistant to acids.
- They can be polished to a mirror finish.
- It is translucent - light can pass through it to the extent of a few millimeters.

Thus, marbles should not be used in high-traffic areas such as the entryways or staircases of public buildings - granite would be much better in those cases. Since it is not acid resistant, you should not use Marble under urinals (urine is acidic), and in kitchens, where lemon juice and other acids are present. But marble is prized for the beauty and richness of its finish; it also *feels* very special underfoot. It is available in a wide variety of colours, mainly light colours.

Sandstone is a sedimentary rock (rock formed by ancient rivers that slowly deposited material on their beds, that built up layer by layer over millions of years). It has the following properties:

- It is abrasion resistant, but not always strong, as it is formed in layers.
- It is usually highly resistant to acids.
- It has a rough finish, and cannot be mirror-polished, as it consists of grains

These properties mean that it is good for decks and external areas because of its anti-slip properties. Since sandstone looks and feels very different from granite and marble, it has become fashionable to use these in boutique stores. Steve Jobs famously saw a bluish-grey sandstone on a trip to Florence, and many years later insisted that that very stone be used in all apple stores because of its 'integrity'. The stone is quarried from an area reserved for apple, cut into tiles, and every piece graded individually for colour tone by master craftsmen. The tiles are then arranged so that pieces with similar colour are placed together, which makes them seem more uniform to the eye.

Common Types of Stones used in Construction

Basalt and Trap

The structure is medium to fine grained and compact. Their colour varies from dark gray to black. Fractures and joints are common. Their weight varies from 18 kN/m^3 to 29 kN/m^3. The compressive strength varies from 200 to 350 N/mm^2. These are igneous rocks. They are used as road metals, aggregates for concrete. They are also used for rubble masonry works for bridge piers, river walls and dams. They are used as pavement.

Granite

Granites are also igneous rocks. The colour varies from light gray to pink. The structure is crystalline, fine to coarse grained. They take polish well. They are hard durable. Specific gravity is from 2.6 to 2.7 and compressive strength is 100 to 250 N/mm^2. They are used primarily for bridge piers, river walls, and for dams. They are used as kerbs and pedestals. The use of granite for monumental and institutional buildings is common. Polished granites are used as table tops, cladding for columns and wall. They are used as coarse aggregates in concrete.

Sand Stone

These are sedimentary rocks, and hence stratified. They consist of quartz and feldspar. They are found in various colours like white, grey, red, buff, brown, yellow and even dark gray. The specific gravity varies from 1.85 to 2.7 and compressive strength varies from 20 to 170 N/mm^2. Its porosity varies from 5 to 25 percent. Weathering of rocks renders it unsuitable as building stone. It is desirable to use sand stones with silica cement for heavy structures, if necessary. They are used for masonry work, for dams, bridge piers and river walls.

Slate

These are metamorphic rocks. They are composed of quartz, mica and clay minerals. The structure is fine grained. They split along the planes of original bedding easily. The colour varies from dark gray, greenish gray, purple gray to black. The specific gravity is 2.6 to 2.7. Compressive strength varies from 100 to 200 N/mm^2. They are used as roofing tiles, slabs, pavements etc.

Laterite

It is a metamorphic rock. It is having porous and sponges structure. It contains high percentage of iron oxide. Its colour may be brownish, red, yellow, brown and grey. Its specific gravity is 1.85 and compressive strength varies from 1.9 to 2.3 N/mm^2. It can be easily quarried in blocks. With seasoning it gains strength. When used as building stone, its outer surface should be plastered.

Marble

This is a metamorphic rock. It can take good polish. It is available in different pleasing colours like white and pink. Its specific gravity is 2.65 and compressive strength is 70–75 N/mm^2. It is used for facing and ornamental works. It is used for columns, flooring, steps etc.

Gneiss

It is a metamorphic rock. It is having fine to coarse grains. Alternative dark and white bands are common. Light grey, pink, purple, greenish gray and dark grey coloured varieties are available. These stones are not preferred because of deleterious constituents present in it. They may be used in minor constructions. However hard varieties may be used for buildings. The specific gravity varies from 2.5 to 3.0 and crushing strength varies from 50 to 200 N/mm2.

Quartzite

Quartzites are metamorphic rocks. The structure is fine to coarse grained and often granular and branded. They are available in different colours like white, gray, yellowish. Quartz is the chief constituent with feldspar and mica in small quantities. The specific gravity varies from 2.55 to 2.65. Crushing strength varies from 50 to 300 N/mm^2. They are used as building blocks and slabs. They are also used as aggregates for concrete.

Stone Masonry and its Types

Stone masonry has been use in many places from years Where stones are easily available, random rubble work will be cheaper than plastered brickwork.

Where the groundwater level is high, and bricks are liable to deteriorate, random rubber work is preferred even today for foundation work.

The sizes of stones used in stone masonry work depend on the type of masonry. The stones should not be larger than what can be handled and placed by one person.

For ashlar work, it is specified that the length should not exceed three times its height.

The breadth should not be greater than three-fourths of the thickness of the wall or less than 15 cm.

The height can be up to 30 cm.

Generally, in random rubble work, larger stones are used on the faces and smaller ones at the hearting. We should be able to identify the various types of stonework as the labor and payment depend on the final appearance of the work. Each of them is specified for different situations.

For example; whereas ordinary random rubble work can be specified in foundation, backside compound walls or garden walls.

Ashlar work will be more fitting in monumental buildings or even as main front walls of buildings and roadside compound walls.

Tools for Stone Masonry

Tools for stone masonry depend on the region where the work is carried out and the workmen carrying out the work.

But following are some of the common tools used in stone masonry:

1. Chisel.
2. Iron Hammer.
3. Mallet.
4. Pick.
5. Spalling Hammer.
6. Claw Tool.

7. Pitching Tool.
8. Jumpers.
9. Saw.
10. Gads.
11. Wedge and Feathers.

Types of Stone Masonry

Stone masonry can be classified broadly into two main groups:

Rubble Masonry

Rubble masonry is made of stone as obtained from the quarry with very little or no dressing.

Types of Rubble Masonry

There are many types of rubble masonry, but the more commonly known ones are the following three according to the CPWD Specification 77:

1. Random rubble masonry.
2. Coursed rubble masonry of the first sort.
3. Coursed rubble masonry of the second sort.

Random Rubble Masonry

Random rubble masonry is the first variety of stone masonry.

Stones are arranged at random in Random Rubble Masonry.

The minimum thickness of random rubble work that can be constructed with great care is 225 mm (9″) and with ease 300 mm (1 ft).

The offsets to be provided in rubble masonry construction have to be 75 mm (3 inches) on either side.

Work for footings has to be adjusted to these sites.

Random rubble masonry is the roughest type of stone masonry.

The stones used are quarried in such a manner and of such sizes so that they can be lifted and placed by hand.

It is only hammer-dressed on the face, the side, and the beds.

The bushing in the face should not project more than 40 mm on an exposed face and 10 mm on a face to be plastered.

Its appearance will be as shown in figure.

The mortar for this masonry is 1:5 up to 1:8.

The mortar for plaster should be 1:3 or 1:4.

The stones are laid on their natural bed on a full even bed of mortar.

Every stone is carefully fitted to the adjacent stone so as to form neat and close joints.

Stones should be wetted before being placed.

Stonework should be brought to a level at window sills and roof level with concrete made of one part of mortar not leaner than that used in the masonry and two parts of graded stone of 20 mm nominal size.

Face stones, stones placed on the faces of the wall, should extend and bond well into the backing.

They should not be less than 125 mm in height and should be hammer-dressed.

These should break joints as much as possible.

The hearting or interior tilling of the wall should consist of rubble stones which may be of any shape but not less than 125 mm in size.

These are laid carefully and hammered down with a wooden mallet into the position and solidly bedded in mortar.

Chips and spalls of stones are used, if needed, to avoid thick mortar beds.

The hearting should be nearly level with the facing. Where the masonry of one part has to be delayed, the work is sloped at an angle, not more than 45 degrees.

Toothing of stonework for joins is not recommended.

A sufficient number of bond stones which are long enough to extend the full thickness of the wall as specified in Figure. should be used in random rubble work.

Joints should be fully packed with mortar and chips.

Face joints should not be more than 20 mm thick. The face can be left as built, pointed or plastered.

If pointing or plastering is to be done, the joints should be raked to a minimum depth of 20 mm by a raking tool (as in brickwork), when the mortar is still green.

Single scaffolding with one set of vertical legs is allowed for the construction of these walls of adequate thickness.

Such holes are later filled with a properly sized stone or with cement concrete 1:3:6 with 20 mm aggregates.

The walls are to be cured for a minimum period of seven days, the fresh work being protected from rain and sun.

Their appearance (plain and pointed masonry) will be as shown in figure.

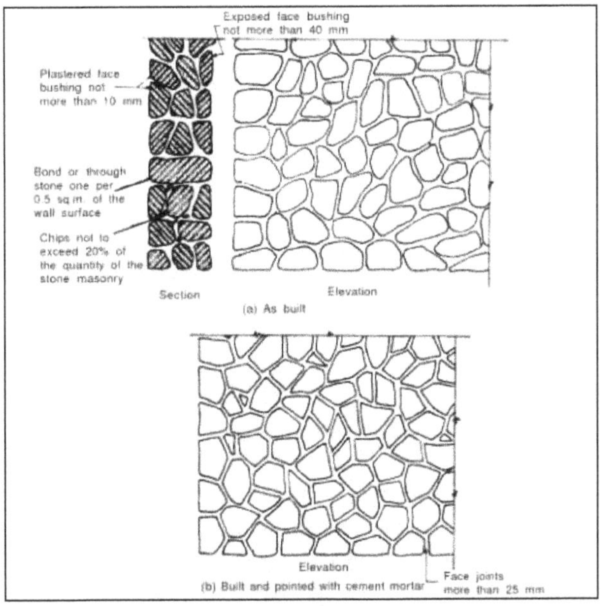

Random rubble masonry.

Coursed Rubble Masonry of the First Sort

This is the second type of rubble masonry and is built in courses (not random).

In this type of work, the stones on both faces are hammer-dressed on all the beds and joints so as to give them approximately rectangular block shape.

This type of rubble masonry should be squared on all joints and beds, and the bed joints shall be rough chisel dressed for at least 80 mm back from the face.

Stones are set in regular courses. The height of the course should not be less than 150 mm.

Work on the interior face is to be the same as on the exterior face.

Similarly, the side joints are dressed for at least 40 mm so that no portion of the dressed surface is more than 6 mm from a straight edge placed on it.

The hammer-dressed stones should have rough tooling for a minimum width of 25 mm along the four exposed edges of the face of the stone.

The bushing on the face should not project more than 40 mm on an exposed face and 10 mm on a face to be plastered.

The percentage of stone chips used should not exceed 10% of the total stones used for the masonry.

In general, the work is carried out using the same mortar as in random rubble work.

Its appearance when finished will be as shown in figure.

The face joints should not be more than 10 mm thick.

Coursed Rubble Masonry of the Second Sort

This is the third type of rubble masonry and is also built in courses.

It is inferior to the coursed rubble masonry of the first sort but will look similar.

No portion of the dressed surface should be more than 10 mm from a straight edge placed on it.

In this type of work, the face joints should not exceed 20 mm in thickness (10 mm in the first sort).

Thus, the joints are allowed to be thicker than those in the first sort.

Its appearance is shown in figure.

This type of work is more expensive than random rubble work but cheaper than course rubble of the first sort.

The percentage of chips used is not to exceed 15% of the quantity of stones in the masonry.

Coursed rubble masonry.

Bond Stones in Rubble Work

One of the important features of rubble masonry work is the use of bond stones.

These are long selected stones placed from front to back of thin walls or from outside to the interior of thick walls as shown in figure.

They hold together the masonry walls transversely.

If single stones of sufficient length are available, a pair of stones each penetrating three-fourths of the thickness of the wall with a minimum overlap of 150 mm.

Or a set of stones each overlapping 150 mm can be used as bond stones.

If separate bond stones of ample lengths are not available or the available stones are porous to develop damp penetration, concrete bond stones made from 1 : 3 : 6 concrete mix and of the specified length can be used as bond stones.

In many specifications, the use of a certain number of bond stones is mandatory.

The recommended specification is that it should not be spaced at more than 1200 mm horizontally and 600 mm vertically.

Ashlar Masonry

Ashlar masonry is also known as cut stone masonry.

Ashlar masonry is laid with carefully-dressed stones and will look like brickwork in stone.

Though ashlar work, if properly constructed, is very strong and beautiful, however, it is very expensive.

Ashlar masonry is divided into the following types:

i. Plain ashlar masonry (rough-tooled and fine-tooled).

ii. Chamfered (also called punched) ashlar masonry (rough-tooled and fine-tooled).

We have seen that the difference between rubble and ashlar masonry is that;

In ashlar masonry, every stone must be cut to the required size and shape to give truly vertical and horizontal joints.

No point on the faces should vary more than about 1 mm when tested with a 60 cm straight edge.

Horizontal lines should not vary more than 3 mm and vertical lines more than 6 mm.

This dressing makes ashlar masonry costlier than rubble masonry.

In short, the completed work will look like "brickwork" in stone.

The two types of ashlar masonry are: plain and chamfered.

They can be either rough tooled or fine tooled.

Types of Ashlar Masonry

1. Plain ashlar masonry (rough tooled and fine tooled).
2. Chamfered or punched ashlar masonry (rough and fine tooled).

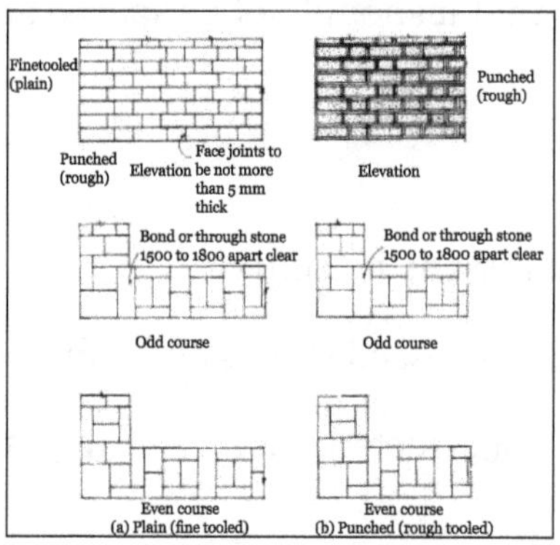

Ashlar stone masonry.

Both are similar in all respects except that the chamfered ashlar masonry will have joints beveled at a 45° angle to a depth of 25 mm.

Rough tooled ashlar masonry will have a chisel dressing 2.5 mm wide all around the edges and shall be roughly tooled between the drafts so that the dressed surface will not be more than 3 mm from a straight edge placed over it.

Its appearance will be as shown in figure.

In the fine tooled work the face will be fine tooled.

In constructing ashlar work, the stones should be covered with gunny bags before ropes are passed over it for lifting so that the stone edges are not damaged.

It should be noted that in the most dressed stonework, the stones in each course are laid as alternate headers and stretchers as shown in figure.

The headers on alternate courses should be approximately in the middle of the stretchers above and below.

Ashlar Facing with Backing of Brickwork (Composite Masonry)

This type of work is executed in many places for its beauty.

In these works, the face stone of ashlar in each course is laid as alternate courses of headers and stretchers unless stated otherwise.

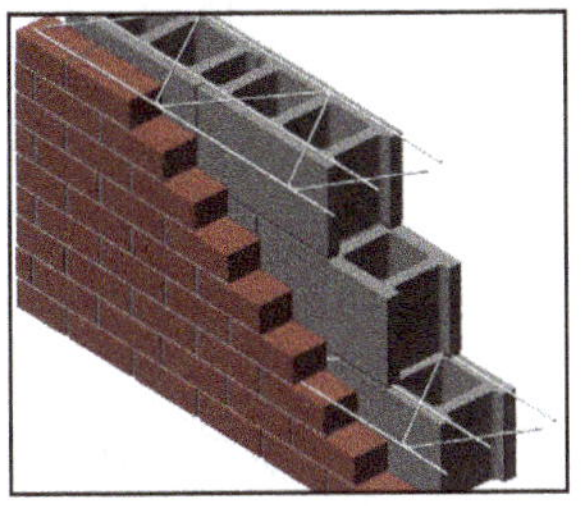

Composite masonry.

Face stones and bond stones must break joints in the face for at least half the height of the standard course, and bond stones should be carefully maintained throughout.

Work on all the connected walls should be carried out at one level throughout, but where breaks are unavoidable, the joint must be made in good long steps so as to prevent the development of cracks.

Pros and Cons of Building a Stone House

Pros

- Sustainability – Stone is a natural construction material, it does not require any other resource to make it. This alone makes it very environment-friendly. It does not also require any use of chemicals to forge because it can be sourced almost anywhere.
- Durable – Stone isresistant to water, fire, wind, and other natural factors that typically affect a house. It does not mold and it is not prone to termites or majority of pests that attack a structure. Basically, stone is damage free.
- Low maintenance – A stone house does not require extensive maintenance even over time. In the long run, it only requires minimal cleaning, and its appearance and quality will still be the same.
- Promotes comfort living – If properly insulated, stone helps the efficient regulation of temperature in a home, keeping you relatively warm during winters and well ventilated during summers.
- Aesthetic appeal – Stones come in many different designs, and there are several ways to work on it. It's only a matter of getting picking the right stone material that will fit with your home design requirements.

Cons

- Expensive – About the most challenging factor to face when building a stone house is the cost of the stone material and its construction.
- Expert contractor – Building a stone house requires highly specialized skills. Expert contractors do not come cheap too.

- Longer construction process – The process of selecting the best stone for your particular home requirement is of utmost importance. Stone is a heavy construction material, it requires special care and handling. Precision is essential when setting it too.

Concrete

Concrete is a hardened building material created by combining a chemically inert mineral aggregate (usually sand, gravel, or crushed stone), a binder (natural or synthetic cement), chemical additives, and water. Although people commonly use the word "cement" as a synonym for concrete, the terms in fact denote different substances: cement, which encompasses a wide variety of fine-ground powders that harden when mixed with water, represents only one of several components in modern concrete. As concrete dries, it acquires a stone-like consistency that renders it ideal for constructing roads, bridges, water supply and sewage systems, factories, airports, railroads, waterways, mass transit systems, and other structures that comprise a substantial portion of the U.S. wealth., building such facilities is in itself one of the nation's largest industries and represents about 10 percent of the gross national product. Over $4 billion worth of hydraulic cement, a variety which hardens under water, is produced annually in the United States for use in $20 billion worth of concrete construction. The value of all cement-based structures in the United States is in the trillions of dollars—roughly commensurate with the anticipated cost of repairing those structures over the next twenty years.

The words cement and concrete are both of Latin origin, reflecting the likelihood that the ancient Romans were the first to use the substances. Many examples of Roman concrete construction remain in the countries that encircle the Mediterranean, where Roman builders had access to numerous natural cement deposits. Natural cement consists mainly of lime, derived from limestone and often combined with volcanic ash. It formed the basis of most civil engineering until the eighteenth century, when the first synthetic cements were developed.

The earliest manmade cement, called hydraulic lime, was developed in 1756, when an English engineer named John Smeaton needed a strong material to rebuild the Eddystone lighthouse off the coast of Devon. Although the Romans had used hydraulic cement, the formula was lost from the collapse of their empire in the fifth century A.D. until Smeaton reinvented it. During the early nineteenth century several other Englishmen contributed to the refinement of synthetic cement, most notably Joseph Aspdin and Isaac Charles Johnson. In 1824 Aspdin took out a patent on a synthetic blend of limestone and clay which he called Portland cement because it resembled limestone quarried on the English Isle of Portland. However, Aspdin's product was not as strong as that produced in 1850 by Johnson, whose formula served as the basis of the Portland cement that is still widely used today. Concrete made with Portland cement is considered superior to that made with natural cement because it is stronger, more durable, and of more consistent quality., Portland cement is made by mixing calcareous (consisting mostly of calcium carbonate) material such as limestone with silica-, alumina-, and iron oxide-containing materials. These substances are then burned until they fuse together, and the resulting admixture, or clinker, is ground to form Portland cement.

Although Portland cement quickly displaced natural cement in Europe, concrete technology in the United States lagged considerably behind. In America, natural cement rock was first discovered during the early 1800s, when it was used to build the Erie Canal. The construction of such inland waterways led to the establishment of a number of American companies producing natural cement. However, because of Portland cement's greater strength, many construction engineers preferred to order it from Europe, despite the additional time and expense involved. Thomas Edison was very interested in Portland cement and even cast phonograph cabinets of the material. When United States industry figured out how to make Portland cement during the early 1870s, the production of natural cement in America began to decline.

After the refinement of Portland cement, the next major innovation in concrete technology occurred during the late nineteenth century, when reinforced concrete was invented. While concrete easily resists compression, it does not tolerate tension well, and this weakness meant that it could not be used to build structures—like bridges or buildings with arches—that would be subject to bending action. French and English engineers first rectified this deficiency during the 1850s by embedding steel bars in those portions of a concrete structure subject to tensile stress. Although the concrete itself is not strengthened, structures built of reinforced concrete can better withstand bending, and the technique was used internationally by the early twentieth century.

Another form of strengthened concrete, prestressed concrete, was issued a U.S. patent in 1888. However, it was not widely used until World War II, when several large docks and bridges that utilized it were constructed. Rather than reinforcing a highly stressed portion of a concrete structure with steel, engineers could now compress a section of concrete before they subjected it to stress, thereby increasing its ability to withstand tension.

Today, different types of concrete are categorized according to their method of installation. Ready- or pre-mixed concrete is batched and mixed at a central plant before it is delivered to a site. Because this type of concrete is sometimes transported in an agitator truck, it is also known as transit-mixed concrete. Shrink-mixed concrete is partially mixed at the central plant, and its mixing is then completed en route to the site.

Raw Materials

Structural concrete normally contains one part cement to two parts fine mineral aggregate to four parts coarse mineral aggregate, though these proportions are often varied to achieve the strength and flexibility required in a particular setting. In addition, concrete contains a wide range of chemicals that imbue it with the characteristics desired for specific applications. Portland cement, the kind most often used in concrete, is made from a combination of a calcareous material (usually limestone) and of silica and alumina found as clay or shale. In lesser amounts, it can also contain iron oxide and magnesia. Aggregates, which comprise 75 percent of concrete by volume, improve the formation and flow of cement paste and enhance the structural performance of concrete. Fine grade comprises particles up to. 20 of an inch (five millimeters) in size, while coarse grade includes particles from. 20 to. 79 of an inch (20 millimeters). For massive construction, aggregate particle size can exceed 1.50 inches (38 millimeters).

Aggregates can also be classified according to the type of rock they consist of: basalt, flint, and granite, among others. Another type of aggregate is *pozzolana,* a siliceous and aluminous material often derived from volcanic ash. Reacting chemically with limestone and moisture, it forms the calcium silicate hydrates that are the basis of cement. Pozzolana is commonly added to Portland cement paste to enhance its densification. One type of volcanic mineral, an aluminum silicate, has been combined with siliceous minerals to form a composite that reduces weight and improves bonding between concrete and steel surfaces. Its applications have included precast concrete shapes and asphalt/concrete pavement for highways. Fly ash, a coal-burning power plant byproduct that contains an aluminosilicate and small amounts of lime, is also being tested as a possible pozzolanic material for cement. Combining fly ash with lime (CaO) in a hydrothermal process (one that uses hot water under pressure) also produces cement.

A wide range of chemicals are added to cement to act as plasticizers, superplasticizers, accelerators, dispersants, and water-reducing agents. Called admixtures, these additives can be used to increase the workability of a cement mixture still in the nonset state, the strength of cement after application, and the material's water tightness. Further, they can decrease the amount of water necessary to obtain workability and the amount of cement needed to create strong concrete. Accelerators, which reduce setting time, include calcium chloride or aluminum sulfate and other acidic materials. Plasticizing or superplasticizing agents increase the fluidity of the fresh cement mix with the same water/cement ratio, thereby improving the workability of the mix as well as its ease of placement. Typical plasticizers include polycarboxylic acid materials; superplasticizers are sulphanated melamine formaldehyde or sulphanated naphthalene formaldehyde condensates. Setretarders, another type of admixture, are used to delay the setting of concrete. These include soluble zinc salts, soluble borates, and carbohydrate-based materials. Gas forming admixtures, powdered zinc or aluminum in combination with calcium hydroxide or hydrogen peroxide, are used to form aerated concrete by generating hydrogen or oxygen bubbles that become entrapped in the cement mix.

Cement is considered a brittle material; in other words, it fractures easily. Thus, many additives have been developed to increase the tensile strength of concrete. One way is to combine polymeric materials such as polyvinyl alcohol, polyacrylamide, or hydroxypropyl methyl cellulose with the cement, producing what is sometimes known as macro-defect-free cement. Another method entails adding fibers made of stainless steel, glass, or carbon. These fibers can be short, in a strand, sheet, non-woven fabric or woven fabric form. Typically, such fiber represents only about one percent of the volume of fiber-reinforced concrete.

The Manufacturing Process

The manufacture of concrete is fairly simple. First, the cement (usually Portland cement) is prepared. Next, the other ingredients—aggregates (such as sand or gravel), admixtures (chemical additives), any necessary fibers, and water—are mixed together with the cement to form concrete. The concrete is then shipped to the work site and placed, compacted, and cured.

Preparing Portland Cement

- The limestone, silica, and alumina that make up Portland cement are dry ground into a very fine powder, mixed together in predetermined proportions, preheated, and calcined (heated to a high temperature that will burn off impurities without fusing the ingredients). Next the material is burned in a large rotary kiln at 2,550 degrees Fahrenheit (1,400 degrees Celsius). At this temperature, the material partially fuses into a substance known as clinker. A modern kiln can produce as much as 6,200 tons of clinker a day.

- The clinker is then cooled and ground to a fine powder in a tube or ball mill. A ball mill is a rotating drum filled with steel balls of different sizes (depending on the desired fineness of the cement) that crush and grind the clinker. Gypsum is added during the grinding process. The final composition consists of several compounds: tricalcium silicate, dicalcium silicate, tricalcium aluminate, and tetracalcium aluminoferrite.

Mixing

- The cement is then mixed with the other ingredients: aggregates (sand, gravel, or crushed stone), admixtures, fibers, and water. Aggregates are pre-blended or added at the ready-mix concrete plant under normal operating conditions. The mixing operation uses rotation or stirring to coat the surface of the aggregate with cement paste and to blend the other ingredients uniformly. A variety of batch or continuous mixers are used.
- Fibers, if desired, can be added by a variety of methods including direct spraying, premixing, impregnating, or hand laying-up. Silica fume is often used as a dispersing or densifying agent.

Transport to Work Site

- Once the concrete mixture is ready, it is transported to the work site. There are many methods of transporting concrete, including wheelbarrows, buckets, belt conveyors, special trucks, and pumping. Pumping transports large quantities of concrete over large distances through pipelines using a system consisting of a hopper, a pump, and the pipes. Pumps come in several types—the horizontal piston pump with semi-rotary valves and small portable pumps called squeeze pumps. A vacuum provides a continuous flow of concrete, with two rotating rollers squeezing a flexible pipe to move the concrete into the delivery pipe.

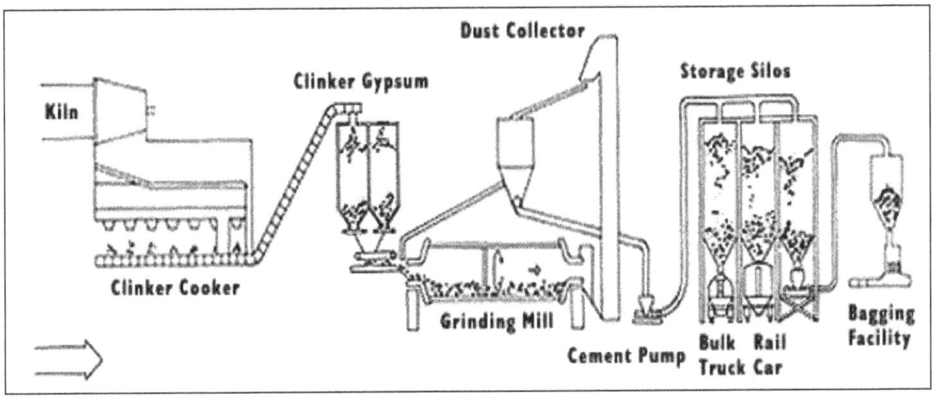

The first step in making concrete is to prepare the cement. One type of cement, Pordand cement, is considered superior to natural cement because it is stronger, more durable, and of a more consistent quality. To make it, the raw materials are crushed and ground into a fine powder and mixed together. Next, the material undergoes two heating steps—calcining and burning. In calcining, the materials are heated to a high temperature but do not fuse together. In burning, however, the materials partially fuse together, forming a substance known as "clinker." The clinker is then ground in a ball mill—a rotating steel drum filled with steel balls that pulverize the material.

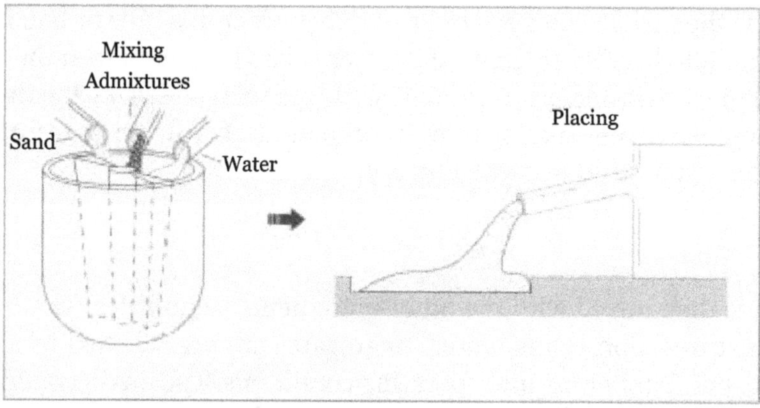

After the Portland cement is prepared, it is mixed with aggregates such as sand or gravel, admixtures, fibers, and water. Next, it is transfered to the work site and placed. During placing, segregation of the various ingredients must be avoided so that full compaction—elimination of air bubbles—can be achieved.

Placing and Compacting

- Once at the site, the concrete must be placed and compacted. These two operations are performed almost simultaneously. Placing must be done so that segregation of the various ingredients is avoided and full compaction—with all air bubbles eliminated—can be achieved. Whether chutes or buggies are used, position is important in achieving these goals. The rates of placing and of compaction should be equal; the latter is usually accomplished using internal or external vibrators. An internal vibrator uses a poker housing a motor-driven shaft. When the poker is inserted into the concrete, controlled vibration occurs to compact the concrete. External vibrators are used for precast or thin in situ sections having a shape or thickness unsuitable for internal vibrators. These type of vibrators are rigidly clamped to the formwork, which rests on an elastic support. Both the form and the concrete are vibrated. Vibrating tables are also used, where a table produces vertical vibration by using two shafts rotating in opposite directions.

Curing

- Once it is placed and compacted, the concrete must cured before it is finished to make sure that it doesn't dry too quickly. Concrete's strength is influenced by its moisture level during the hardening process: as the cement solidifies, the concrete shrinks. If site constraints prevent the concrete from contracting, tensile stresses will develop, weakening the concrete. To minimize this problem, concrete must be kept damp during the several days it requires to set and harden.

Quality Control

Concrete manufacturers expect their raw material suppliers to supply a consistent, uniform product. At the cement production factory, the proportions of the various raw materials that go into cement must be checked to achieve a consistent kiln feed, and samples of the mix are frequently examined using X-ray fluorescence analysis.

The strength of concrete is probably the most important property that must be tested to comply with specifications. To achieve the desired strength, workers must carefully control the manufacturing process, which they normally do by using statistical process control. The American Standard of Testing Materials and other organizations have developed a variety of methods for testing strength. Quality control charts are widely used by the suppliers of ready-mixed concrete and by the engineer on site to continually assess the strength of concrete. Other properties important for compliance include cement content, water/cement ratio, and workability, and standard test methods have been developed for these as well.

Grade of Concrete

Grade of concrete denotes its strength required for construction. For example, M30 grade signifies that compressive strength required for construction is 30MPa. The first letter in grade "M" is the mix and 30 is the required strength in MPa.

Based on various lab tests, grade of concrete is presented in Mix Proportions. For example, for M30 grade, the mix proportion can be 1:1:2, where 1 is the ratio of cement, 1 is the ratio of sand and 2 is the ratio of coarse aggregate based on volume or weight of materials.

The strength is measured with concrete cube or cylinders by civil engineers at construction site. Cube or cylinders are made during casting of structural member and after hardening it is cured for 28 days. Then compressive strength test is conducted to find the strength.

Regular grades of concrete are M15, M20, M25 etc. For plain cement concrete works, generally M15 is used. For reinforced concrete construction minimum M20 grade of concrete are used.

Concrete Grade	Mix Ratio	Compressive Strength	
		MPa (N/mm2)	psi
Normal Grade of Concrete			
M5	1 : 5 : 10	5 MPa	725 psi
M7.5	1 : 4 : 8	7.5 MPa	1087 psi
M10	1 : 3 : 6	10 MPa	1450 psi
M15	1 : 2 : 4	15 MPa	2175 psi
M20	1 : 1.5 : 3	20 MPa	2900 psi
Standard Grade of Concrete			
M25	1 : 1 : 2	25 MPa	3625 psi
M30	Design Mix	30 MPa	4350 psi
M35	Design Mix	35 MPa	5075 psi
M40	Design Mix	40 MPa	5800 psi

M45	Design Mix	45 MPa	6525 psi
High Strength Concrete Grades			
M50	Design Mix	50 MPa	7250 psi
M55	Design Mix	55 MPa	7975 psi
M60	Design Mix	60 MPa	8700 psi
M65	Design Mix	65 MPa	9425 psi
M70	Design Mix	70 MPa	10150 psi

Types of Concrete

1. Plain or Ordinary Concrete.
2. Lightweight Concrete.
3. High-Density Concrete.
4. Reinforced Concrete.
5. Precast Concrete.
6. Prestressed Concrete.
7. Air Entrained Concrete.
8. Glass Concrete.
9. Rapid Hardening Concrete.
10. Asphalt Concrete.
11. Lime Concrete.
12. Roller Compacted Concrete.
13. Stamped Concrete.
14. Pumped Concrete.
15. Vacuum Concrete.
16. Permeable Concrete.
17. Shotcrete.
18. Ready-mix Concrete.
19. Self-Consolidated Concrete.
20. Fiber Reinforced Concrete.

21. Fly Ash Concrete.
22. High Strength Concrete.
23. Silica Fume Concrete.
24. Polymer Concrete.
25. Ferro Cement Concrete.
26. Pre-packed Concrete.

Plain or Ordinary Concrete

It is one of the most commonly used types of concrete. In this type of concrete, the essential constituents are cement, sand and coarse aggregates designed, and mixed with a specified quantity of water.

The ratio of essential constituents may be varied within wide limits. A very commonly used mix design, commonly known as *Nominal Mix Design,* is 1:2:4.

Plain concrete is mostly used in the construction of pavements and in buildings, where very high tensile strength is not required. It is also used in the construction of Dams.

Among the most important properties of ordinary concrete, the following may be mentioned.

1. Density: 2200 – 2500 Kg/meter.cube.
2. Compressive Strength: 200 – 500 Kg/centimeter.square.
3. Tensile Strength: 50 – 100 Kg/centimeter.square.
4. Durability: Very satisfactory.

Lightweight Concrete

Any types of concrete having a density less than *1920 Kg/m³* is classed as *lightweight concrete.*

Various types of aggregates that are used in the manufacturing of lightweight concrete include natural materials like pumice and scoria, artificial materials like expanded shales and clays and processed materials like perlite and vermiculite.

The single important property of lightweight concrete is its very low thermal conductivity.

For example: Thermal conductivity – the k value, for plain concrete may be as high as 10-12. But the thermal conductivity of Lightweight concrete is about 0.3.

Lightweight Concretes are used, depending upon their composition, for thermal insulation, for protecting steel structures, they are also used in long span bridge decks, and even as building blocks.

Aerated Concrete is a variety of extremely lightweight concrete (density 480-800 Kg/m^3). This is obtained by using cement, sand, and powdered fuel ash as constituents.

High-Density Concrete

This type of concrete is also called *heavy weight concrete.* In this concrete type, the density varies between 3000-4000 Kg/m^3.

These types of concrete are prepared by using high density crushed rocks as coarse aggregates. Among such materials, barytes is the most commonly used material, which has a specific gravity of 4.5.

They are mostly used in atomic power plants and other similar structures because it provides good protection from all type of radiations.

Reinforced Concrete

It is also called RCC (Reinforced Cement Concrete). In this concrete type, steel in various forms is used as reinforcement to give very high tensile strength.

In fact, it is because of the combined action of plain concrete (having high compressive strength) and steel (having high tensile strength).

The steel reinforcement is cast in the form of rods, bars, meshes, and all conceivable shapes.

Every care is taken to ensure the maximum bond between the reinforcement and the concrete during the setting and hardening process.

Thus, the resulting material (RCC) is capable of bearing all types of stress in any type of construction. RCC is the most important concrete type.

Precast Concrete

This term refers to numerous types of concrete shapes that are cast into molds either in a factory or at the site.

However, they are not used in construction until they completely set and hardened in a controlled condition.

Some of the examples of Precast Concrete are; *precast poles, fence posts, concrete lintels, staircase units, concrete blocks, and cast stones*, etc.

These structural and decorative members are prepared in a well-equipped place where all arrangements are made for;

1. Perfect proportioning of the ingredients of concrete.
2. Thorough mixing of the cement, aggregates, and water to obtain the mix of the desired design and consistency.
3. Careful handling during transport and placement in the perfect design molds.
4. Perfect curing, under the controlled conditions of temperature and humidity. Even steam curing is used to obtain precast products having high strength in much less time.
5. The latest trend in the construction industry is to shift more and more to prefabricated concrete units in building construction.

Prestressed Concrete

It is a special type of reinforced concrete in which the reinforcement bars are tensioned before being embedded in the concrete.

Such tensioned wires are held firm at each end while the concrete mix is placed. The result is that when the concrete sets and hardens, the whole concrete members, so the cast is put into compression.

This sort of arrangement makes the lower section of the reinforced concrete also stronger against tension, which is the principal cause of the development of tension cracks in un-tensioned reinforced concrete.

Since pre-stressing involves the use of jacks and tensioning equipment, the pre-stressed concrete is also cast in the factories.

Some of its advantages are following:

1. The potential compressive strength of concrete gets considerably increased.
2. The risk of development of tension cracks in the lower sections of beams is considerably reduced.

3. The resistance to shear is greatly reduced. This eliminates the necessity of stirrups to a great extent.
4. Lighter members can be used than the un-tensioned (normal) reinforced-concrete.
5. The prestressed concrete is greatly favored in the construction of:
    - Bridges.
    - Long span Roofs.
    - Most structures with the heavy dead load.

Air Entrained Concrete

It is a specially prepared plain concrete in which air is entrained in the form of thousands of uniformly distributed particles.

The Volume of air thus, entrained may range between 3-6 percent of the concrete.

The air entrainment is achieved by adding a small quantity of foaming or gas-forming agents at the mixing stage.

Fatty acids, fatty alcohols, and resins are some common air entraining agents.

Air entrained concrete is more resistant to;

1. Scaling.
2. Deterioration due to freezing and thawing.
3. Abrasion.

Glass Concrete

When the recycled glass is used as an aggregate in the concrete, this type of concrete is known as Glass Concrete.

They provide better thermal insulation and also have a great appealing look as compared to other types.

Rapid Hardening Concrete

This type of concrete is mostly used in underwater construction and in repairing of roads. Because its hardening time is very less, it can be hardened in just a few hours.

They are also used in building construction, where the work should be done fast.

Asphalt Concrete

Asphalt concrete is a combination of aggregates and asphalt. It is also known as Asphalt. They are vastly used in the highways, airports, as well as in the embankments.

They can be hardened in just an hour. That is the reason for its vast usage in roads.

Lime Concrete

In this type of concrete, lime is used as a binding material with the aggregates. Before the invention of cement, the mostly used concrete was lime concrete.

In today's age, Lime concrete is also used in floors, domes, etc.

Roller Compacted Concrete

This concrete is mostly used as a filling material. They don't have a better strength value. They are lean concrete and are compacted with the help of heavy means, like rollers.

Very less amount of cement is used in this type of concrete.

Stamped Concrete

They are ordinary concrete with some little differences and are mostly used for architectural purposes.

A stamp of different shape and design placed on the concrete structures when they are in their plastic state to acquire an appealing look design.

Pigments are used for color purposes of different types to give it a more realistic and appealing look.

Pumped Concrete

Pumped concrete is used for high rise buildings where concrete conveyance other than the pump is not easy task.

They are made workable enough for an easy conveyance. Fines material are used for a better supply. The more, the finer material, the easy will be the discharge.

The pump used for conveyance purposes is made from rigid or flexible materials to discharge the concrete easily.

Vacuum Concrete

In this type, more quantity of water is added to the concrete mix, and then the mixture is poured into the formwork.

The excess water is then removed from the concrete with the help of a vacuum pump. That is why it is called vacuum concrete.

This technique is used to attain the strength of concrete early. It will attain the compressive strength within the period of 10 days as compared to 28 days of ordinary concrete.

Permeable Concrete

Permeable concrete is prepared in such a manner that the water can be passed in it. They have about 15 to 20% voids so that the water can pass in it.

They are used in those areas where stormwater issues persist.

Shotcrete

Shotcrete is a concrete prepared in the same manner as ordinary, but the difference is that they are placed differently.

They are placed with the help of higher air pressure through nozzles. The benefit of this technique is that the compaction and placing of concrete will be done simultaneously.

Ready Mix Concrete

This concrete type is prepared in concrete plants and or transported by the help of truck mounted transit mixtures.

Once they are reached at the site then, there is no further treatment necessary.

The plant location will be at an adjustable location so that the concrete can be supplied before the setting time ca be started.

Self Consolidated Concrete

These types of concrete are compacted by its own weight, mean by the process of consolidation. There is no need for using a vibrator or doing manual compaction.

The workability of concrete is always high in this type. That is the reason it is also known as flowing concrete.

Fiber Reinforced Concrete (FRC)

It is that type of concrete in which steel fibers 10 to 20 microns in diameter and 10 to 50 mm in length is used.

Fiber increases resilience, tensile strength, flexibility, and other qualities.

The fibers may be of different materials like steel, polymer, glass, carbon, or even natural fibers like coconut fiber.

Some type of fibers react with the cement; special care should be taken while using them. It has been used mostly as overlays for pavements in bridges, airports, and over industrial floors.

Fiber reinforced concrete can also be used in places where increased resistance to cracking is required.

Fly Ash Concrete

Concrete using fly ash is called fly ash concrete. Fly ash is obtained from coals. Fly ash can be used to replace fine aggregates or cement or to replace partially both.

Up to 30 percent replacement of fine aggregates and 20 percent replacement of cement have been reported.

Fly ash improves workability in the fresh concrete and durability and strength in hardened concrete.

The particles of fly ash should be finer than of cement particles.

High Strength Concrete

High-strength concrete is concrete with strength over 40 N/mm^2. It is also known as High-performance concrete (HPC).

High-performance concrete is used to achieve some special properties in concrete like high strength, low shrinkage, self compaction, high fire resistance, etc.

Normally, the strength of such concrete should be over 60 N/mm^2 (Strengths up to 80 N/mm^2 have been reported).

The materials used in the HPC are the following:

1. Cement,
2. Coarse and fine aggregates of the required quality,
3. Water,
4. Supplementary cementing materials like silica fume, fly ash, blast furnace slag, etc.
5. Superplasticizers (high water reducing agents),
6. Air entraining agents (optional).

Silica Fume Concrete

Silica fume is a byproduct of silica which is very finely divided in the industry. Concrete in which silica fume is used is called "silica fume concrete."

The typical concrete with normal water-cement ratio always has micro-pores, which limits the strength of regular concrete.

Silica fumes consist of very fine particles (actually, 6 times finer than of cement particles).

Hence, if it is added to the concrete mix, the minute pore spaces can be reduced, resulting in high-strength concrete.

Silica fume is also a pozzolana which will contribute to the strength. Thus, silica fume, along with superplasticizers is a necessary component of high-performance and high-strength concrete.

Polymer Concrete

Polymerization is a process of conversion of monomers into polymers. In normal concrete, you should have seen that micro-pores cannot be avoided.

The impregnation of monomer into these pores and subsequent polymerization is the technique that has been developed recently to reduce the porosity of the concrete and to improve its strength and other properties.

The following are the four types of polymer concrete materials available at present:

Polymer Impregnated Concrete (PIC)

This concrete type is ordinary-cured concrete, which is dried in an oven, and the air in the open cells is removed by vacuum.

A low viscosity monomer is introduced into these spaces, which is then polymerized by application of chemical action or heat or by using radiation.

Polymer Portland Cement Concrete (PPCC)

This types of concrete are made by mixing a monomer along with the mixing of aggregates, cement, and water. However, the concrete obtained by this method is not as strong as the impregnated-type.

Polymer Concrete (PC)

In this type of concrete, instead of cement, the polymer is used with the aggregates. It is not a true concrete as used in civil engineering terminology.

Partially-Impregnated and Surface-Coated Polymer Concrete

This name is self-explanatory. All these materials at present are mostly used in building construction only to improve durability or repairs works.

Ferro Cement Concrete

Ferro cement concrete should not be confused with fiber concrete. Ferro cement consists of closely spaced wire-meshes which are impregnated with a rich mix of cement mortar.

Usually, 0.5 to 1.0 mm diameter steel wires are formed into meshes.

Mortar 1:2 to 1:3 with a water-cement ratio of 0.4 to 0.45 is poured into the form-work with fabricated steel by using layers of the wire mesh.

The steel content of this concrete will be as high as 300 to 500 kg/m^3 of mortar. As the material consists of a large percentage of steel, it has high ductility and tensile strength.

The material was developed in 1940 by the Italian architect P. L. Nervi to build a large number of pleasing structural forms.

Pre-Packed Concrete

Generally, concrete is prepared by mixing different ingredients.

However, it is also possible to pack some of the ingredients (coarse aggregate) in the form-work and then fill the pores with specially prepared cement-sand grout so that it will fill all the pores and form a concrete mass.

Pre-packed concrete is used in special situations such as where a large volume of concrete (like a large machine block foundation) has to be concreted without construction joints.

One of the advantages of pre-packed concrete is that it has very little shrinkage.

Concrete Frame Structures

Concrete frame structures are a very common - or perhaps the most common- type of modern building internationally. As the name suggests, this type of building consists of a frame or skeleton of concrete. Horizontal members of this frame are called *beams*, and vertical members are called *columns*. Humans walk on flat planes of concrete called *slabs*. Of these, the column is the most important, as it is the primary load-carrying element of the building. If you damage a beam or slab in a building, this will affect only one floor, but damage to a column could bring down the entire building.

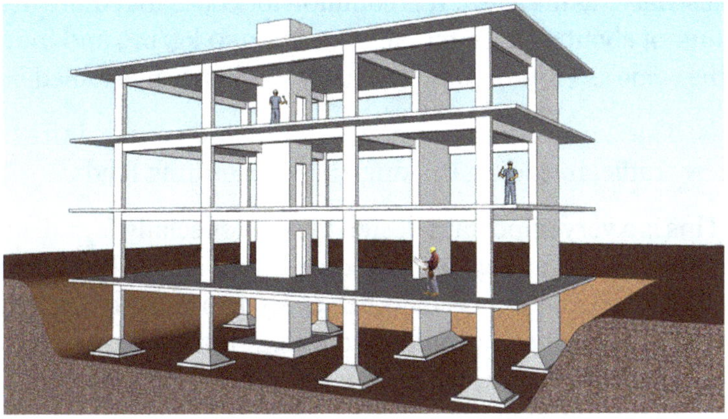

When we say concrete in the building trade, we actually mean reinforced concrete. Its full name is reinforced cement concrete, or RCC. RCC is concrete that contains steel bars, called reinforcement bars, or rebars. This combination works very well, as concrete is very strong in compression, easy to produce at site, and inexpensive, and steel is very very strong in tension. To make reinforced concrete, one first makes a mould, called *formwork*, that will contain the liquid concrete and give it the form and shape we need. Then one looks at the structural engineer's drawings and places in the steel reinforcement bars, and ties them in place using wire. The tied steel is called a *reinforcement cage*, because it is shaped like one.

Once the steel is in place, one can start to prepare the concrete, by mixing cement, sand, stone chips in a range of sizes, and water in a cement mixer, and pouring in the liquid concrete into the formwork tilll exactly the right level is reached. The concrete will become hard in a matter of hours, but takes a month to reach its full strength. Therefore it is usually propped up until that period. During this time the concrete must be *cured*, or supplied with water on its surface, which it needs for the chemical reactions within to proceed properly.

Working out the exact 'recipe', or proportions of each ingredient is a science in itself. It is called *concrete mix design*. A good mix designer will start with the properties that are desired in the mix, then take many factors into account, and work out a detailed mix design. A site engineer will often order a different type of mix for a different purpose.

So the structure is actually a connected frame of members, each of which are firmly connected to each other. In engineering parlance, these connections are called *moment connections*, which means that the two members are firmly connected to each other. There are other types of connections, including *hinged connections*, which are used in steel structures, but concrete frame structures have moment connections in 99.9% of cases. This frame becomes very strong, and must resist the various loads that act on a building during its life. These loads include:

- Dead loads: The downwards force on the building coming from the weight of the building itself, including the structural elements, walls, facades, and the like.
- Live loads: The downwards force on the building coming from the expected weight of the occupants and their possessions, including furniture, books, and so on. Normally these loads are specified in building codes and structural engineers must design buildings to carry these or greater loads. These loads will vary with the use of the space, for example, whether it is residential, office, industrial to name a few. It is common for codes to require live loads for residential to be a minimum of about 200 kg/m^2, offices to be 250 kg/m^2, and industrial to be 1000 kg/m2, which is the same as 1T/m^2. These live loads are sometimes called *imposed loads*.
- Dynamic loads: These occur commonly in bridges and similar infrastructure, and are the loads created by traffic, including braking and accelerating loads.
- Wind Loads: This is a very important design factor, especially for tall buildings, or buildings with large surface area. Buildings are designed not to resist the everyday wind conditions, but extreme conditions that may occur once every 100 years or so. These are called *design windspeeds*, and are specified in building codes. A building can commonly be required to resist a wind force of 150 kg/m^2, which can be a very significant force when multiplied by the surface area of the building.
- Earthquake loads: In an earthquake, the ground vigorously shakes the building both horizontally and vertically, rather like a bucking horse shakes a rider in the sport of rodeo. This can cause the building to fall apart. The heavier the building, the greater the force on it. Its important to note that both wind and earthquake impose *horizontal forces* on the building, unlike the gravity forces it normally resists, which are vertical in direction.

The concrete frame rests on foundations, which transfer the forces - from the building and on the building - to the ground.

Some other important components of concrete frame structures are:

Shear Walls are important structural elements in high-rise buildings. Shear walls are essentially very large columns - they could easily measure 400mm thick by 3m long - making them appear like walls rather than columns. Their function in a building is to help take care of *horizontal forces* on buildings like wind and earthquake loads. Normally, buildings are subject to vertical loads - gravity. Shear walls also carry vertical loads. It is important to understand that they only work for horizontal loads in one direction - the axis of the long dimension of the wall. These are usually not required in low-rise structures.

Elevator Shafts are vertical boxes in which the elevators move up and down - normally each elevator is enclosed in its own concrete box. These shafts are also very good structural elements, helping to resist horizontal loads, and also carrying vertical loads.

Walls in Concrete Frame Buildings

Concrete frame structures are strong and economical. Hence almost any walling materials can be used with them. The heavier options include masonry walls of brick, concrete block, or stone. The lighter options include drywall partitions made of light steel or wood studs covered with sheeting boards. The former are used when strong, secure, and sound-proof enclosures are required, and the latter when quick, flexible lightweight partitions are needed.

When brick or concrete blocks are used, it is common to plaster the entire surface - brick and concrete - with a cement plaster to form a hard, long-lasting finish.

Cladding of Concrete Frame Structures

Concrete frame buildings can be clad with any kind of cladding material. Common cladding materials are glass, aluminum panels, stone sheets, and ceramic facades. Since these structures can be designed for heavy loading, one could even clad them in solid masonry walls of brick or stone.

Concrete Construction Processes

Batching Concrete

The process of measuring ingredients or materials to prepare concrete mix is known as batching of concrete. Batching can be done by two methods, volume batching and weight batching. Batching should be done properly to get quality concrete mix.

Methods of Batching Concrete

Volume Batching

In volume batching, materials are measured on the basis of volume. It is less precise method of batching Measurement boxes or gauge boxes of known volume are used to measure materials. Cement is taken in the form of bags, where volume of one bag of cement (50 kg) is taken as 35 liters.

Volume of Gauge box used is made equal to the volume of one bag of cement which is 35 liters or multiple thereof. Gauge boxes are generally deeper and contains narrow top surface and they are made of timber or steel or iron. Volumes of different sized fine aggregate and coarse aggregate are measured individually by these gauge boxes.

Water is measured using water meter or water cans of known volume are used. To make 1:1:2 ratio concrete mix according to volume batching, one should take one bag of cement (35 liters) , 1 gauge box of fine aggregate (35 liters) and 2 gauge boxes of fine aggregate (70 liters). If the water-cement ratio is 0.5, then half of the volume of cement which is 25 liters of water should be taken.

Gauge Box or Measuring Box.

Weight Batching

In this method, Materials are measured on the basis of weight. It is accurate method of batching. Weigh batchers or other types of weighing equipment are used to measure weight of materials. Cement, fine aggregate, coarse aggregate and water are taken by weighing. Weigh batchers used are available in two types namely mechanical weigh batcher and electronic weigh batchers.

In mechanical weigh batchers, weights are measured using spring and dial gauge arrangement and it is widely used equipment in weigh batching. In electronic weigh batchers, electronic scales and load cells supported by hoppers are used to measure the weight of ingredients of concrete. Weigh batchers available are may be Manual or semi-automatic or fully automatic . Manual type is used for small concrete production job while other two types are used for large concrete production.

Semi-automatic Weigh Batcher.

In case of semi-automatic weigh batching, aggregate container gates are lifted manually and it is automatically closed after reaching required quantity in the weighing machine.

In fully automatic weigh batcher, all the process will be done automatically. The benefit of this type equipment is, it also measures the moisture content present in the aggregate and corrects the required quantity of water-cement with respect to moisture content of aggregates.

To prepare 1:1:2 concrete mix using weigh batching, measured quantity of materials are 50 kg of cement, 50 kg of fine aggregate and 100 kg of coarse aggregate.

Automatic Weigh Batcher.

Weight Batching vs. Volume Batching

Weigh batching is most precise method of batching than volume batching since it is difficult to find the exact volume of granular materials because of their voids. Moisture content presence in the aggregate should also be considered while batching. In case of fully automatic weigh batching it can be considered but volume batching is not suitable in such cases.

Compressive strength of same concrete mix at 7 days and 28 days is higher for weigh batching concrete mix than volume batching concrete mix. Weigh batching concrete mix gives medium to very high slump while for the same mix proportion in volume batching, slump differs from low to high. Volume batching does not require skilled workers but weigh batching. Selection of batching method depends upon size of project, concrete production rate.

Concrete Batching Plant.

Mixing Concrete

Making concrete is very easy, but proper mixing of concrete ingredients is the most important because it affects the quality of concrete.

A Well-Mixed Concrete should Fulfill the following Requirements:

- Concrete should be uniform in colour.
- Homogeneous mixing of all concrete materials like cement, fine aggregate, coarse aggregate and water.

- Cement paste should cover all the surface of the aggregate.
- Segregation should not occur after the mixing of concrete.

When it comes to Mixing Concrete, there are three Options to Choose from:

- Hand mixing (Mixing concrete without a mixer).
- Machine mixing (Mixing concrete with a mixer).
- Ready mix concrete.

Hand Mixing (Mixing Concrete without a Mixer)

Hand mixing is the process of mixing the various materials of concrete manually. Mixing concrete without a mixer is used only for small works. Mixing of materials shall be done on masonry platform or flat iron sheet plates.

The hand mixing concrete may be done as detailed below:

- Spread the measured quantity of sand on the platform, and then cement shall be dumped on the sand.
- The sand and cement shall be mixed thoroughly with the help of shovels in the dry state.
- The measured amount of coarse aggregate shall be spread out, and the mixture of sand cement spread on it and mixed properly.
- Depression is made at the centre of the mixed materials.
- Add 75% of the required quantity of water in the depression and mix with the help of shovels.
- Add the remaining amount of water and continue the mixing process till a uniform colour and consistency of concrete is obtained. Time of mixing concrete should not exceed 3 minutes.
- The mixing platform should be washed at the end of the day's work.

Machine Mixing (Mixing Concrete with a Mixer)

While hand mixing is a quite good process to ensure the quality of concrete, in larger projects where large masses of concrete are required, it is neither feasible nor advisable. Hand mixing also doesn't ensure the consistent homogeneity of concrete. Also, the demands for short mixing time and consistency, homogeneous quality of concrete are increasing day by day, which cannot be satisfied by hand mixing. Machine mixing can satisfy all the demands mentioned above. Nowadays various types of concrete mixers are available which are either petrol/diesel or electrically powered.

The machine mixing may be done as detailed below:

- First of all, the concrete mixer should be wetted inside of the drum.
- After that Cement, sand and coarse aggregate shall be placed in the portable concrete mixer in required proportion.
- The dry materials shall be mixed in the mixing machine. After this, correct quantity of water shall be added gradually while the machine is in motion.
- You must mix concrete for minimum two minutes after all materials are in the drum.
- If there is segregation, after unloading from the mixer, the concrete shall be remixed.

Ready Mix Concrete

In India, traditionally concrete has been produced on site with the primitive mixer machine equipments and use of labour force. Ready Mixed Concrete (RMC) is the concrete which is produced in the factory or in a batching plant and delivered in a ready-to-use manner. The quality of the resulting concrete is better than the site-mixed concrete.

Hand mixing is a cheaper method of mixing concrete but recommended only for very small works. Proper mixing of material is achieved in machine mixing only and hence used for big projects. Ready mix concrete requires less time as compared to site mixing (hand or machine mixing) and quality of concrete is also better than the site mixing.

Transportation and Placement of Concrete

After mixing, concrete should be transported and placed at site as quickly as possible without segregation, drying, etc. Hence transportation of fresh concrete is very important operation. Depending upon the type of work and equipment available various methods of transporting of concrete can be employed. The method selected should be suitable for use with the consistency of concrete employed.

It is important that the concrete be handled without segregation of the constituent materials. This condition may easily occur if special attention is not given to its prevention. Segregation occurs because concrete is not a homogenous combination, but a mixture of materials differing in size and specific gravity. Thus as soon as the concrete is discharged from the mixer, internal as well as external forces start acting to separate the dissimilar constituents. If over wet concrete is confined in restricting forms, the coarser and heavier particles tend to settle and finer and lighter materials tend to rise.

Thus the equipment for conveying concrete from mixer to form should be such as to result in uniform concrete of the desired consistency at the form. The method of conveying should not require wetter concrete than is required at form for placing nor it should produce segregation or excessive drying or stiffening etc.

To avoid segregation, when concrete is to be transferred from one conveyance to another, use of hoppers, baffles and short vertical drops should be used through a pipe to the centre of the receiving container as shown in figure. This belief is not correct that segregation produced during conveyance can be rectified by compaction. Segregation should be prevented rather than corrected after its occurrence.

Figure shows correct and incorrect method of filling hoppers or buckets. It shows the control of separation of concrete at the end of conveyer belts, while figure shows control of separation at the end of chutes. In this case a 60 cms long pipe is provided at the end of the chute. This figure shows the correct and incorrect methods of concrete discharge. Figure shows placing of concrete on a sloping surface. A baffle and drop at the end of chute will avoid separation and concrete remains on slope.

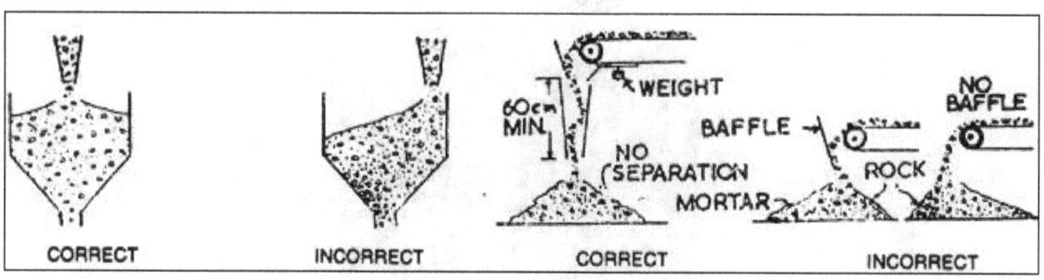

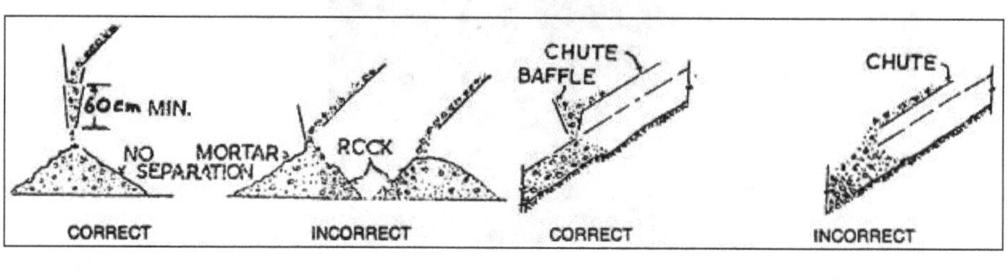

Methods of Conveyance for Transporting Concrete

For moving concrete from mixer to the form following methods may be employed:

1. Pan method
2. Wheel barrows
3. Dumpers
4. Lorries
5. Mono rail transporters
6. Hoists
7. Crane
8. Skips and buckets
9. Chutes
10. Conveyer belts
11. Pneumatic pressure equipment, and
12. Pumps.

Ready Mixed Concrete

Concrete available for placing from a central plant is known as ready mixed concrete or pre mixed concrete.

Advantages of Ready Mixed Concrete

Following are the advantages of ready mixed concrete:

1. This type of concrete is particularly useful on congested sites.
2. For road construction where little space is available for mixing plant and for stocking aggregate etc.
3. Ready mixed concrete can be made under better control than on site. Actually a factory type control on all operations is possible on the central mixing plant. This is the greatest advantage of ready mixed concrete.

   Ready-made concrete can be considered as a factory made product almost comparable with steel. Hence a great deal of uncertainty and variability associated with the production of concrete on many sites is removed.
4. The use of ready mixed concrete is advantageous when only small quantities of concrete is required or concrete is to be placed at intervals.

Disadvantages

The main disadvantage of ready mixed concrete is its high cost. It is costlier by about 10 to 15% than concrete mixed at site.

Types of Ready Mix Concrete

Ready mix concrete can be classified into two groups as follows:

1. Central mixed concrete.
2. Transit mixed or truck mixed concrete.

Central Mixed Concrete

In this case the mixing is done at a central plant and the mixed concrete is transported in a special type of truck known as an agitator truck. The truck (agitator truck) revolves slowly while moving. The agitating container is conical in shape having smaller blades than mixer. The capacity of the container is 1.25 m^3 to 2.0 m^3. However the height of this container is more than a mixer. The container is put upright in the truck and to check the evaporation of water from the concrete an air tight stopper is placed on the top of the container.

The segregation and stiffening of concrete is checked by revolving the container. The revolving of the container is also done by the engine of the truck. The rate of revolving of the container depends upon the speed of the lorry varying 2 to 6 revolutions per minutes. However 2 to 3 revolutions per minute are quite sufficient.

Advantages of Agitating Trucks

1. To revolve the container, very less amount of power is needed.
2. During even in one hour journey, no ill effect on the properties of concrete is developed.

Disadvantage

The main disadvantage of this method is that these agitating trucks cannot be used for any other purpose. Hence this method is uneconomical.

Transit Mixed Concrete

In this method the materials are batched at a central plant and the dry mix is put in mixers kept in trucks. The mix is mixed either during transit to the site or immediately prior to the concrete being discharged at the site. The speed of rotation is between 4 to 16 revolutions per minute. ASTM has laid down the limit of 300 revolutions for both agitating and mixing, and concrete must be laid within 90 minutes of mixing. The capacity of mixer may vary from 0.75 m to 4.5 m^3.

Advantages

1. Transit mixing permits a longer haul and is less vulnerable in case of delay.

2. As the water is added to mix near the site of work, dislocation of the work does not take place.
3. The discharge of concrete is done by reversing the direction of the revolving mixer rather than tipping the truck.
4. These mixers have small revolving tubing, which can spread the concrete on a sufficient wide area. This operation makes the concrete spreading operation economical.

Disadvantages

1. Transportation by this method is limited to an area within a radius of 9 to 15 km.
2. These trucks can be used only for this purpose.
3. The moist aggregate should not remain in contact with cement for more than 90 minutes.

The agitating differs from mixing solely by the speed of rotation of the mixer. The agitating speed is between 2 and 6 revolutions per minute, whereas mixing speed varies from 4 to 16 revolutions per minute. It may be noted that the speed of mixing affects the rate of stiffening, while the total number of revolution controls the uniformity of mixing.

Causes of Blockage during Concrete Mix

Generally Blockages take Place Due to the following Causes:

1. Unsuitability of the concrete mix i.e. concrete mix is not designed properly.
2. Deficiencies in pipe line and joints.
3. Careless use of hose end.
4. Error of operator.

Sometimes use of admixtures, particularly accelerating admixture and high grade cement, high temperature may cause blockage. If continuous pumping is not done there are more chances of blockage.

If the pipe line is not cleaned properly after its previous use there are more chances of blockage. The use of un-cleaned, worn out hoses, too many and too sharp bends, use of worn out joints are also other causes of blockages.

Thus to avoid blockages, the hose must be well lubricated. Sufficient quantity of lubricating grout should be used to cover the entire length of pipe line before pumping concrete. The flexible rubber end hose should be handled with care. Careless handling may cause blockage.

Location of the Blockage

The location of blockage can be determined by tapping the pipe line and observing the sound emitted from the pipe line.

Remedial Measures of Blockage

1. A minor blockage may be cleared by forward and reverse pumping. However excess pressure should not be applied blindly. It may aggravate the problem.
2. Blockage can be cleared by rodding or using sponge ball pushed by compressed air or water at high pressure.
3. Sometimes shortening the pipe line reduces pressure and on restarting the pumping, blockage is removed.

Safety Precautions during Transportation of Concrete

Usually following precautions should be adopted during transportation of concrete:

1. Places where Lorries are to be discharged should have strong support for tipping the Lorries.
2. No person should be allowed to stand at the place where the tipping lorry discharges concrete.
3. To avoid segregation, distance between the bottoms of the skip to the top of surface where concrete is to be placed should not be more than one metre.
4. The distance between wagons being moved simultaneously by hand in one direction should not exceed 20 m on level surface and 30 m on slopes.
5. The whole system of pipe line should be checked by using hydraulic pressure 1.5 times the operating pressure before pumping the concrete. Generally water is used for this purpose.
6. The run of wheel barrow should be clean and at least 1.2 m wide.
7. While using hoist, the driver of the hoist should be in a position to see the place of charging and discharging the hoist.

Placement of Concrete

The process of depositing the concrete in its required position is termed as placement of concrete.

When concrete is to be placed at a place it should be deposited as near as possible to its final position, so that when is it to be compacted it does not have to flow appreciably. Areas which will be in contact with concrete should be prepared well before placing concrete. Hence in order to get good finished surfaces, it is essential that surface where concrete is to be laid should be prepared carefully.

Preparation of Surface before Placing Concrete

In order to develop a proper and adequate bond between the base and the concrete placed, it is essential that the base is prepared well before placing the concrete.

Before placing concrete different types of bases should be prepared as discussed below:

Natural Soil

Before laying any concrete layer over a natural soil base, the soil should be properly compacted and leveled. After leveling the coarse grained soil should be made moist for a thickness of 30 cm and loamy and fine grained soils upto 15 cm thickness for 24 hours before laying the concrete layer. The moistening of soil is essential, to check absorption of water from concrete by the soil. However water should not be allowed to stand on the prepared surface.

In Case of Rocky bases

In this case the sides should be kept vertical. All loose particles should be removed and water sprinkled over the base and cleaned before placing the concrete. In this case also water should not be allowed to stand on the surface.

In Case of Brick Soling and W.B.M bases

The base should be made rough with the help of wire brush and cleaned. Loose particles and dust removed, water sprinkled and surface compacted before placing the concrete over it.

In Case of Hardened Concrete base

a. The difference in two operations is 4 hours. The laitance on the surface of old concrete should be removed by rubbing it with wire brush and cleaned with water. All dust and rubbish should be removed by broom before laying the new layer of concrete.

b. Upto 48 hours. The surface is rubbed with steel brush and washed with water and cleaned with the help of the broom. Now 1.5 cm thick layer of cement and sand mortar of similar composition as that of concrete is applied to the surface before new concrete is placed.

c. More than 48 hours old concrete. In this case the hard surface is chiselled or sand blasted and cleaned as above Now a slurry of neat cement is applied on the roughened surface and worked into the interstices after which 1.5 cm thick layer of cement sand mortar of the same composition as that of concrete should be applied before the slurry dries and then new concrete is placed immediately.

Placing of Concrete

Placing and compacting operations are the most critical field activities. Unless care is exercised in this operation a very poor job may result even though concrete prepared is the best. To avoid segregation is most important if it occurs remedial measures should be taken to eliminate it, so that homogeneous concrete is obtained.

When concrete is placed in deep forms, it is usually allowed to drop regardless of height. These results in segregation damage to form and embedded fixtures.

Reinforcement and forms above the level of placement become coated with mortar which may dry before the concrete comes to that level. This condition should be avoided by dropping the concrete into an outside pocket and allowing it to flow over into the form without segregation. Figure shows correct and incorrect method of pouring concrete in deep forms.

Pouring of concrete.

In case of horizontal layers, it should be deposited close to its final location. In case of placing concrete in a slab, it should be dumped into face of previously placed concrete not away from the previously placed concrete. Figure shows a correct and incorrect method of placing concrete in slab with the help of buggies.

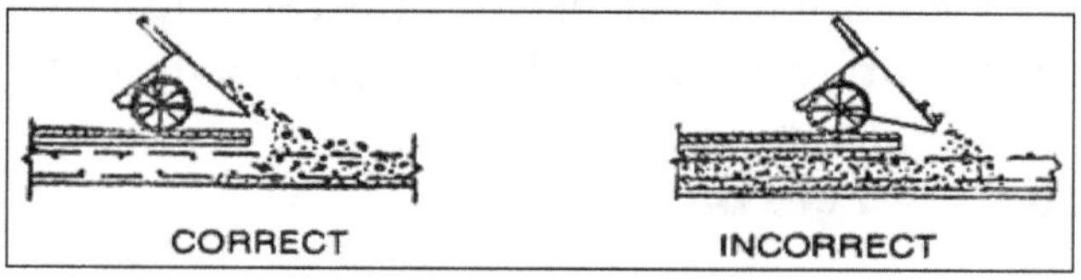

Placing of concrete.

When concrete is to be placed on a slight slope, placing should begin at the lower end of the slope, permitting consolidation of the layer by vibration or any other suitable means. If placement is started at the top of the slope vibration tends to shift it down the slope. Figure shows a correct and incorrect method of placing concrete on a slope.

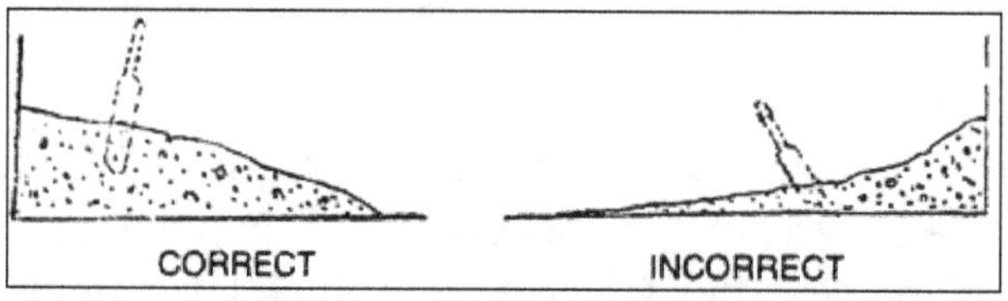

Placing concrete on slopes.

Precautions

Following precautions should be taken while placing concrete:

1. In case of reinforced concrete, thickness of concrete may vary from 15 cms to 30 cms. and 45 cms in case of mass concrete. II should be laid in horizontal and uniform layers.
2. Concrete should not be thrown from a height of more than 1 m to prevent segregation.
3. No person should be allowed to walk over freshly laid concrete.
4. Placement of concrete should be discontinued during heavy downpour i.e., during heavy rain fall period.

5. Before placing new concrete, lataince over the old surface should be removed.
6. In order to prevent formation of irregular surface, concrete should be laid continuously.
7. The alignment of reinforcement and form work should not be disturbed.
8. In order to avoid sticking of concrete to surface of the form work, the internal faces of form work should be oiled or greased.
9. In R.C.C. slabs concreting should start from width wise from one end.
10. In order to produce better bond between concrete and reinforcement, grease, oil, dry cement etc. should be removed from the reinforcement.
11. While pouring concrete in deep forma, concrete should not be thrown from a height.

Compaction of Concrete

The process of compaction of concrete consists essentially of the elimination of entrapped air from the concrete. The oldest means of achieving this aim is by punning with a rod, or by ramming or by vibrating. Thus the object of compacting concrete is to obtain a dense mass of concrete without voids, to get the concrete to surround all reinforcement and to fill all corners.

Compaction of concrete influences the strength, durability, impermeability of concrete to a great extent. Each one percent less compaction reduces the strength of concrete by about 5% on an average. During the process of manufacture of concrete a considerable amount of air is entrapped forming voids in it.

Voids in Concrete

Voids present in concrete in the form of small pores reduce the strength and density of concrete.

Three kinds of voids may be present in the concrete as follows:

1. Water Voids

   These voids are present due to the water in excess of that needed for hydration.

2. Air Voids

   These are either due to insufficient compaction or due to deliberate entrainment of air in the mix. Concrete containing air voids left due to insufficient compaction of the freshly placed concrete lowers the strength very much. Experimental results have shown that 10% air voids left in concrete reduce its strength by more than 50%. Thus freshly laid concrete should be compacted sufficiently so that air voids are completely removed. Voids due to poor compaction can be seen readily when they are at the surface. The effect of air voids is shown in figure.

   Figure shows the position of fresh un-compacted concrete having about 10% air voids. In this condition its compressive strength is found 122.5 kg/cm^2 and weight 2166 kgm^3. Figure shows 5% air voids in the concrete, resulting its strength and weight as 190.5 kg/cm^2 and

2286 kg/m³; while 11.1 (c) shows a fully compacted concrete having its strength and weight as 272.0 kg/cm² and 2410 kg/m³ respectively. This shows the influence of air voids on the strength and weight of concrete. Honey-combed concrete does not develop good bond with reinforcement. Water may penetrate through these voids and corrode the steel.

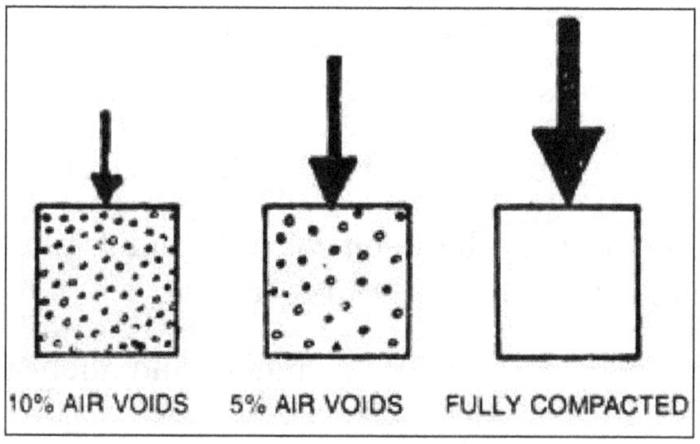

Effect of amount of compaction on stremgth and density of concrete.

Methods of Compaction for Concrete

The method of compaction depends upon its workability, so that is can be fully compacted. In other words the required workability of concrete depends on how much it is to be compacted and by what method.

Usually following methods may be adopted for concrete compaction:

1. Hand Compaction or hand tamping.
2. Mechanical compaction.

Hand Compaction

This method is adopted for members having reinforcement, pavements, narrow and deep members etc. Rammers and iron rods are used for this purpose. Mass concrete should be compacted in layers not more than 30 cms in thickness with templates or light rammers. For compacting reinforced concrete works iron rods are used. In this case the thickness of concrete layers should not be more than 15 cms. The most satisfactory method of compacting concrete properly is to consolidate each layer separately so that its top surface becomes level and fairly smooth before the next layer is placed.

When the rod is inserted vertically into a mass of workable concrete and moved up and down, it causes the concrete to be set in motion similar to that of the rod. This action is known is tamping or punning. When the concrete is set in motion in this way, the air bubbles present in the concrete come to the surface.

Tamping should be carried out for each layer in such a way that concrete is filled in all the corners of the form work. While tamping is carried out care should be taken that the rod should penetrate the full depth of the last layer placed and to some extent into the under lying layer to ensure proper bond between them. Secondly the reinforcement and form work should not be disturbed from their positions.

Tamping of Road Slabs

For compacting a concrete slab such as for roads, floors etc. a steel shod wooden tamping beam or screed is used as shown in figure. It can also ensure the correct level of the surface. The total length of the tamper should be 22.5 cms more than the width of the slab. For compacting the base coarse the tamper should be recessed near the ends by 2.5 cms as shown in figure.

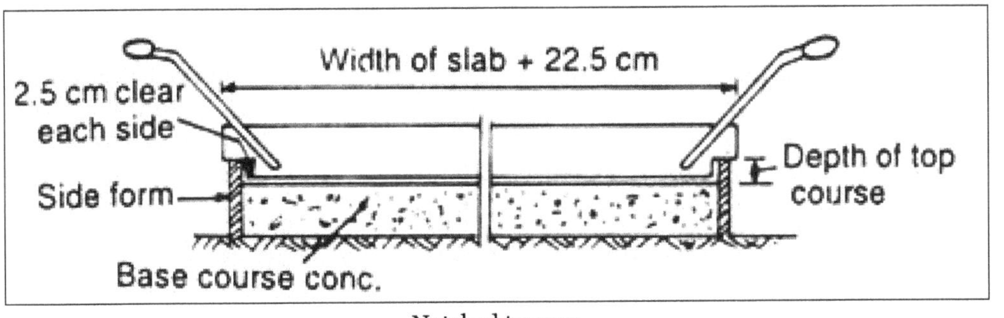

Notched tamper.

The concrete is placed in the form, so that it is deeper than the side forms and the tamper is bumped up and down on the top surface of the concrete. By this actin concrete goes down to the level of the form. During tamping process, any low spots should be filled with concrete sprinkling from a shovel and tamped into the surface. A second run with the tamper is done with a to and fro motion to get the desired surface level.

Finishing Concrete Slabs

The operations adopted for obtaining a true, and uniform concrete surface are called finishing operations. A tamper usually leaves a slightly ridged surface. Thus it needs finishing. Actually the choice of concrete finish depends upon the use of the completed work.

Finishing Operations

For finishing floors and slabs following operations are adopted:

i. Screeding

ii. Floating

iii. Troweling

Screeding

The levelling operation which removes bumps and hollows and gives a true and uniform concrete surface is known as screeding. For this purpose a straight edge is used. It is slightly longer than the section to be finished. The surface is struck off by moving the straight edge back and forth across the top of the surface. The straight edge is advanced forward a very little distance with each movement.

A small quantity of concrete mix should always be kept ahead of the straight edge to fill the voids and to maintain a plane surface. Screeds may be of roller type or vibrator type.

Floating

The operation of removing the irregularities from the surface of the concrete left after screening is known as floating. It helps in leveling the surface and compacting the concrete. Floating is done with the help of 1.5 m long and 20 cms wide wooden floats. To give ease in handling the float, a handle is provided at its centre. Finishing is done by moving the float forward and backward.

Troweling

The final operation of finishing is known as troweling. It is done where smooth surface is desired. Troweling should be done after the evaporation of water from the concrete surface. Spreading of dry cement on wet surface to absorb excess water should not be allowed as it is not a good practice, and will develop cracks and blistering.

Types of Surface Finish

According to the roughness of the surface, the surface finish can be classified as follows:

Brush Finish

Brush finish can only be done after the surface has been floated or belted. It is done with a bamboo broom brushed lightly across the slab after the concrete has set slightly. If a rougher surface is desired, further brushing may be done till the aggregate becomes exposed.

Belt Finish

A belt finish is produced by using a strip of canvas about 15 cms wide and about 1.5 m longer than the width of slab. The belt is laid across the surface and drawn to and fro with rather long strokes and moved forward slowly in the direction of concreting.

Suitability of Concrete Mix

If concrete is to be compacted successfully by hand, then the concrete mix should be of proper proportions and required consistency.

Concrete to be hand compacted should have following slumps as shown in table:

S.no	Type of job	Slump in cms
1.	Heavily reinforced sections . and works where compaction is difficult	10 to 15
2.	Normal reinforced concrete. beams, slabs, columns, walls and footings	7.5 to 10
3.	Road slabs mass concrete, foundations and walls.	2.5 to 7.5

If the concrete is highly deficient in fine aggregate or highly under sanded it may not be possible to compact it, however wet it is made or however much it is tamped. Such a mix would be very harsh and the mortar would tend to separate out from the coarse aggregate causing honeycombing even though thorough tamping may be applied to it.

On the other hand if the mix is over sanded, to make it workable larger amount of water is needed. In such cases excessive maintenance would be formed on the surface after the concrete has been compacted. This laintence being a layer of slurry of cement and sand forms a weak and porous layer on the surface of the lift. Such layers often allow water to leak through construction joints of retaining walls and produce unsightly staining.

Mechanical Compaction

Mixes which are too stiff to consolidate by hand, they can be easily compacted by mechanical compaction. When a concrete mix is vibrated, it behaves as if it is very much wetter than it really is and will flow easily. This method of Compaction has proved very much effective than hand compaction.

In this case the concrete is compacted by vibration, during which the vibrator communicates rapid vibrations to the particles, increasing the fluidity of concrete. Due to vibration, the particles occupy a more stable position and the concrete fills all the space and the air present is forced out to the surface, resulting in dense and durable concrete. The frequency, duration of vibration etc., depends on the conditions of vibration for compacting of the mix.

Advantages of Vibrated Concrete

Following are the advantages of compaction by vibrations:

i. By the use of vibration, stiff mixes can be compacted easily.

ii. For a given cement/aggregate ratio a much stronger and dense concrete can be produced.

iii. Bond between reinforcement and concrete is improved.

iv. Creep and shrinkage is reduced.

v. Laintence is also reduced to some extent.

vi. Speed of placing concrete is increased.

vii. Better finish is obtained.

Types of Vibrators used for Concrete Compaction

Following three types of vibrators usually are used for concrete compaction:

Internal Vibrator

These vibrators are inserted into the concrete and withdrawn after the compaction has been completed.It is also known as immersion, poker or needle vibrator. Usually it is considered to be the most effective type of vibrator as it comes into intimate contact with the concrete.

Though there are different kinds of internal vibrators, but they essentially consist of the following components:

i. A power unit

ii. Flexible tube

iii. Vibrating head

A Power Unit

It can be either a petrol engine or electric supply unit.

Flexible Tube

It is a rubber tube having power connecting device at one end and vibrator head on the other end.

Vibrator Head

It is a straight iron tube having eccentric weight fitted inside it. As the current passes through this tube, the vibratory action is caused by the rotation of the eccentrically weighted shaft at high speed, usually over 7000 r.p.m. which are transmitted to the concrete particles.

The needle diameter may vary from 2.0 cms to 7.5 cms and length from 25 cms to 90 cms. The frequency of vibration may vary upto 12000 cycles per minute. The bigger needle is used in the construction of mass concrete works as dams.

The vibrating head is immersed in the fresh concrete and transmits vibrations to the concrete through the vibrator's body. The vibrating head should be immersed vertically in the concrete and should be kept inside the concrete. While running the vibrator, the bearings should be kept cool and breakdowns should be avoided. As the concrete consolidates, its surface level falls and the surface becomes smooth. Air bubbles may be seen breaking at the surface. When these bubbles cease to appear on the surface, the concrete may be assumed as compacted sufficiently.

The vibrator should be lowered or inserted vertically in the concrete. After the concrete is compacted, the vibrator should be withdrawn slowly at the rate of 7.5 to 8 cm/sec. as this speed will fill the hole caused due to withdrawal of vibrator without any air being entrapped. Now the vibrator is inserted at another point at a distance of 50 cm to 100 cm or 45 cm to 75 cm. However this distance depends on the shape of the form work and amount of reinforcement. This distance should be kept uniform. For concrete of workability of 0.78 to 0.85 compacting factor the vibrator should be spaced at points 35 to 90 cms apart.

The internal vibrator neither should be allowed to tough the inside faces of the form work, as it may damage the surface nor the reinforcement as the displacement of reinforcement will upset all calculations.

The frequency of vibration varies from 3500 to 12000 cycles per minute. Frequency between 3500-5000 cycles per minute has been suggested as a desirable minimum, with an acceleration of not less than 4 g, but recent studies found favour for frequencies from 4000-7000 cycles per minute.

Time of Vibration

The time of compaction by the internal vibrator depends on the consistency of the mix. Usually from 5 to 30 seconds compaction is sufficient depending upon the consistency of the mix, but for some mixes upto 120 seconds compaction is required. The actual completion of compaction can be judged by the appearance of the surface of the concrete, which should neither contain excessive mortar nor it should be honey combed.

Some researchers have correlated the time of compaction by internal vibrators with the consistency of the mix and the frequency of the vibrator. For complete compaction of concrete having 1.25 cm slump required 90 seconds vibration at 4000 r.p.m., 45 seconds at 5000 r.p.m. and 25 seconds at 6000 r.p.m.

When concrete is placed in layers the vibrator should be immersed through the entire depth of the freshly deposited concrete and into the layer below it, if this is still plastic or can be brought again into a plastic condition. In this way a plane of weakness at the junction of the two layers can be avoided.

For layers more than 50 cms, the vibrator may not be fully effective in expelling air from the lower part of the layer. As stated above, the vibrator should remain vertical while in use. Figure shows a correct and incorrect position of the vibrators.

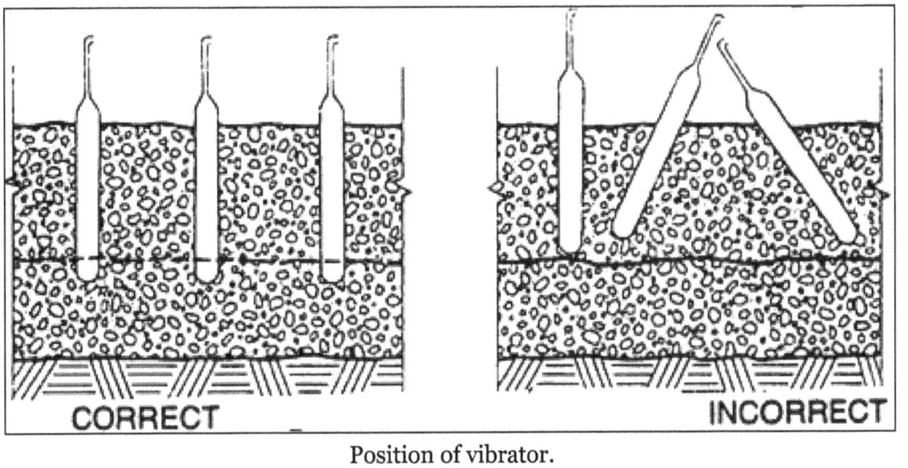

Position of vibrator.

Precautions to be Observed while using Internal Vibrator

1. In order to prevent any leakage of mortar, the form work should be as tight as possible.
2. In order to avoid air being trapped, the thickness of the concrete lift should not be less than 15 cm.
3. For lifts more than 50 cm, internal vibrators should not be used, but upto 45 cm lift, the vibrator should be allowed to penetrate into the previously compacted concrete below to obtain a proper bond between the two lifts or layers. If required the vibrating unit can be immersed completely upto the bottom of the form.
4. Vibrations should be controlled carefully by immersing the vibrator for 5 to 15 seconds at a point. The distance between two immersion points of the vibrator may vary from 45 to 75 cm or 50 to 100 cm.
5. The vibrator should be immersed vertically and should remain in vertical position throughout the operation at that point otherwise the degree of compaction cannot be controlled well.
6. Vibrator should not be used to push concrete laterally in the form works as it will cause segregation of the concrete.

7. The vibrator should be withdrawn slowly usually 7.5 cm to 8.0 cm per second as at this speed of withdrawal, the hole left by vibrator will close fully without any air being trapped.
8. The vibrator should be immersed through the full depth of freshly laid concrete and into the layer below it if the concrete is still plastic or it can be brought into plastic state.
9. The vibrator neither should touch the inside of form work nor the reinforcement.
10. The vibrator should never be kept running for long when it is not in the concrete.
11. The tube should never be bent at a very sharp angle as it will produce unnecessary heavy strain on the internal flexible drive and the outer tube. The tube should be kept as straight as possible.

Sharp bends should be avoided near the junction of the tube and the vibrating head. Internal vibrators are made in sizes down to 20 mm diameter so that they can be used for heavily reinforced sections. These classes of vibrators are found comparatively more efficient as all the work is done directly on the concrete. Though internal vibrators are usually constructed very strong and are capable being used under very rough conditions, however they cannot be expected to work continuously for an indefinite period of time without wearing.

The bearings inside the vibrating head are subject to severe wearing conditions and are likely to become over heated. This can be prevented by immersing the full length of vibrating head into the concrete.

Form Vibrator or External Vibrators

These vibrators are clamped on the side of the form work.These vibrators are used for compacting concrete where internal vibrators cannot be used such as for very heavily reinforced or very narrow sections or precast units etc. In such conditions vibration is applied through the form work by clamping these vibrators on the shuttering.

The vibrators are rigidly clamped to the form work resting on an elastic support, so that both the form and concrete are vibrated. In this case a considerable proportion of work done is consumed in vibrating the form work, resulting in low efficiency of the system. In order to check the leakage of cement mortar the form work should be strong and tight.

Principle of Working

The principle of an external vibrator is same as that of an internal vibrator, but the frequency is usually between 3000 and 6000 cycles of vibration per minute. Bureau of reclamation has recommended at least 8000 cycles per minute.

When an external vibrator is used, concrete has to be placed in layers of suitable depth as air cannot be expelled through a great thickness of concrete. The position of vibrator has to be changed as concreting progresses. They should not be placed more than 90 cms apart in any direction.

The form vibrator consists of an electric motor with an unbalanced weight. This weight causes the unit to vibrate when it is rotated by the motor. These vibrators should be fixed on the form work firmly and should not be placed more than 1 m apart.

Kango Hammer

It is another form of external vibrator. It is an electric hammer used on the outside of the vertical shuttering. This type of vibrator is useful for vibrating those parts of shuttering where it would be difficult to fix the ordinary clamp on the vibrator or where the vibrator is required to move continuously.

Vibrating Tables

These are also external vibrators and are suitable for pre-cast concrete work. It provides a reliable means of compaction of precast concrete and has the advantage of offering uniform vibration.

Surface Vibrators

These vibrators are placed on the top of the concrete surface and concrete compacted.These vibrators are used where a wide horizontal surface occurs such as in dams and very thick walls. Usually such vibrators are used for roads and floors. Though a large number of different types of surface vibrators exist, but pan type vibrator is generally used for floors and foundation works. It consists of a flat pan or tray with handles at each end and a vibrating unit mounted in the middle. The vibratory unit may have a petrol engine or electric motor or may be pneumatic type. This type of vibrator is moved across the surface of the concrete by the handles.

Vibrating Screeds

External vibrator type unit is often used for surface vibration by mounting it on a screed or tamper. Vibrating screeds may have more than one unit of such vibrators. These screeds can be used for compacting concrete layers not more than 20 cm in thickness.

Compaction by Pressure and Jolting

This method has proved very effective for compacting very dry concrete. This method is used for compacting hollow blocks, cavity blocks and solid concrete blocks. The stiff concrete is vibrated, pressed and also jolted. With the combined effect of vibration, pressure and jolt, the stiff concrete gets compacted to a dense form to give good strength and volume stability. By applying great pressure, concrete of low water/cement ratio can be compacted to yield very high strength.

Compaction by Spinning

This is one of recent methods of compaction of concrete. This method of compaction is employed for the production of concrete pipes. When plastic concrete is spun at a very high speed, it gets compacted by the action of centrifugal force. Patented pipes as “Hume pipes”, “Spun pipes” are compacted by this method.

Vibratory Rollers

Compaction of very dry and lean concrete by vibratory rollers is one of the recent developments. Concrete compacted by this method is known as Roller Compaction Concrete. This method of concrete compaction was developed in Japan and spread to USA and other countries mainly for the construction of dams and pavements.

For compaction of dry lean concrete, heavy rollers which vibrate while rolling are used.

Selection of Vibrators for Various Situations

Actually the use of a certain type of vibrators depends upon the situation of work.

Broadly following table gives an indication of the use of the different vibrators:

S.no	Situation where a vibrator used	Type of vibrator
1.	For foundation concrete, columns beams, large sections of mass concrete,	Internal vibrator
2.	For deep beam sections, pre-cast concrete, tunnel lining arches etc.	Form vibrator
3.	Road surfaces, floors of plain and, reinforced concrete	Surface vibrator

Suitability of Mix for Compaction with Vibrator

Concrete to be compacted by vibration, should be designed properly. The consistency of the concrete depends upon the conditions of placing, type of mix, and the efficiency of the vibrator. The slump of such concrete should not be more than 5 cm in any case otherwise segregation of concrete will take place, which should never be allowed to occur.

The value of slump and compacting factors for different works are shown in table:

S.no	Placing condition	Degree of workability	Average size aggregate in mm	Compacting factors	Slump in mm
1.	(i) Narrow sections requiring heavy van ration	Very low	10-20	0.75 - 0.78	0-12
	(ii) Large sections which require normal vibration	Very low	40	0.78	0-25
2.	(i) Normal reinforced sections compacted by vibration	Low	10	0.83-0.85	0-6
	(ii) Heavy sections without reinforcement				6-25
		Low	40	0.85	25-50
3.	(i) Normal reinforced sections compacted without vibration	Average	10-20	0.9-0.92	6-25
					25-50
	(ii) Heavy reinforced section compacted without vibration	Average	40	0.92	50-100
4.	Heavy reinforced section compacted without vibration	High	10-20	0.92-0.95	25-100
					50-100
	Normal sections not suitable to be Compacted by vibration	High	40	0.95	100-175

Comparison of Internal and External Vibrators

1. Internal vibrators are more efficient as all the work is done directly on the concrete. They can be used for concrete layers upto 50 cm thickness.
2. Internal vibrators can be used even for heavily reinforced sections.
3. External vibrators are clamped rigidly to the form work hence a considerable portion of the work done is wasted in vibrating form work.

4. Form work used for external vibrators should be very rigid and strong. This increases the cost of the form work.
5. By the use of external vibrator, excessive cement mortar comes to the surface of the concrete work. This surface being comparatively weak and less pervious deteriorates very soon.

Output of Immersion Vibrators

For light vibrators having a centrifugal force of about 200 kg, the output of compacted concrete may be taken as 3 to 5 m^3 per hour depending upon the consistency of the mix and for heavy type vibrators having a centrifugal force of 450 kg, the output may be taken as 12 to 25 m^3 per hour.

Curing of Concrete

Curing is the maintenance of a satisfactory moisture con- tent and temperature in concrete for a period of time immediately following placing and finishing so that the desired properties may develop. The need for adequate curing of concrete cannot be overemphasized. Curing has a strong influence on the properties of hardened concrete; proper curing will increase durability, strength, watertight- ness, abrasion resistance, volume stability, and resistance to freezing and thawing and deicers. Exposed slab surfaces are especially sensitive to curing as strength development and freeze-thaw resistance of the top surface of a slab can be reduced significantly when curing is defective.

When portland cement is mixed with water, a chemical reaction called hydration takes place. The extent to which this reaction is completed influences the strength and durability of the concrete. Freshly mixed concrete normally contains more water than is required for hydration of the cement; however, excessive loss of water by evaporation can delay or prevent adequate hydration. The surface is particularly susceptible to insufficient hydration because it dries first. If temperatures are favorable, hydration is relatively rapid the first few days after concrete is placed; however, it is important for water to be retained in the concrete during this period, that is, for evaporation to be prevented or substantially reduced.

With proper curing, concrete becomes stronger, more impermeable, and more resistant to stress, abrasion, and freezing and thawing. The improvement is rapid at early ages but continues more slowly thereafter for an indefinite period. Figure shows the strength gain of concrete with age for different moist curing periods and figure shows the relative strength gain of concrete cured at different temperatures.

Curing should begin as soon as the concrete stiffens enough to prevent marring or erosion of the surface. Burlap sprayed with water is an effective method for moist curing.

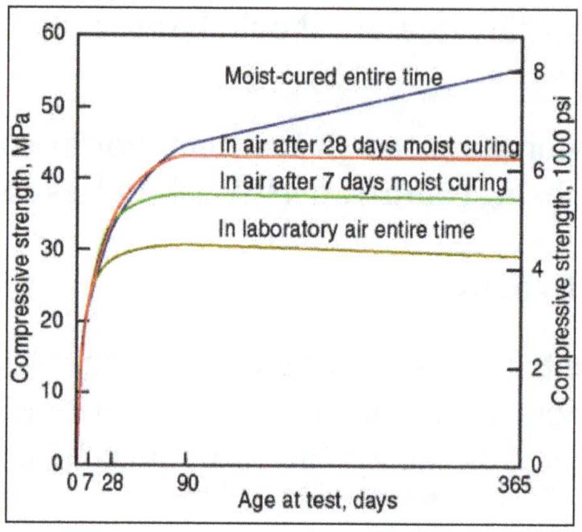

Effect of moist curing time on strength gain of concrete.

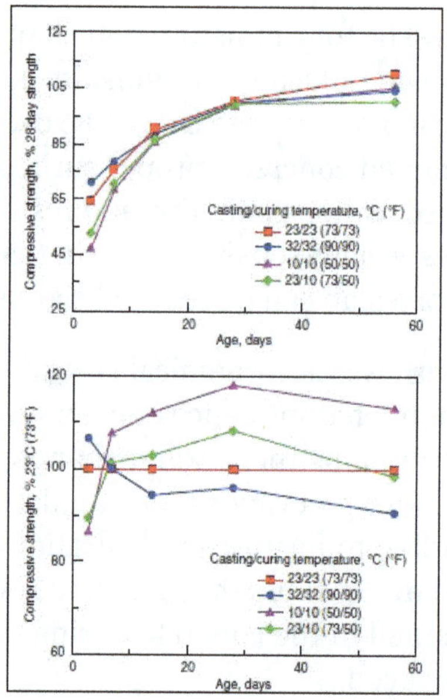

Effect of curing temperature on strength gain (top) relative to 28-day strength and (bottom) relative to the strength of concrete at 23 °C (73 °F).

The most effective method for curing concrete depends on the materials used, method of construction, and the intended use of the hardened concrete. For most jobs, curing generally involves applying curing compounds, or covering the freshly placed and finished concrete with impermeable sheets or wet burlap. In some cases, such as in hot and cold weather, special care using other precautions is needed.

Concrete mixtures with high cement contents and low water-cement ratios (less than 0.40) may require special curing needs. As cement hydrates (chemically combining with water) the internal relative humidity decreases causing the paste to self-desiccate (dry out) if no external water is provided. The paste can self-desiccate to a level where hydration stops. This may influence desired concrete properties, especially if the internal relative humidity drops below 80% within the first seven days. In view of this, membrane-forming curing compounds may not retain enough water in the concrete. Therefore, fogging and wet curing become necessary to maximize hydration. Fogging during and after placing and finishing also helps minimize plastic cracking in concretes with very low water-cement ratios (especially around 0.30 or less).

When moist curing is interrupted, the development of strength continues for a short period and then stops after the concrete's internal relative humidity drops to about 80%. However, if moist curing is resumed, strength development will be reactivated, but the original potential strength may not be achieved. Although it can be done in a laboratory, it is difficult to resaturate concrete in the field. Thus, it is best to moist-cure the concrete continuously from the time it is placed and finished until it has gained sufficient strength, impermeability, and durability.

Loss of water will also cause the concrete to shrink, thus creating tensile stresses within the concrete. If these stresses develop before the concrete has attained adequate tensile strength, surface cracking can result. All exposed surfaces, including exposed edges and joints, must be protected against moisture evaporation.

Hydration proceeds at a much slower rate when the concrete temperature is low. Temperatures below 10 °C (50 °F) are unfavorable for the development of early strength; below 4 °C (40 °F) the development of early strength is greatly retarded; and at or below freezing temperatures, down to -10 °C (14 °F), little or no strength develops.

In recent years, a maturity concept has been introduced to evaluate the development of strength when there is variation in the curing temperature of the concrete. Maturity is the product of the age of the concrete and its average curing temperature above a certain base temperature. It follows that concrete should be protected so that its temperature remains favorable for hydration and moisture is not lost during the early hardening period.

Curing Methods and Materials

Concrete can be kept moist (and in some cases at a favorable temperature) by three curing methods:

1. Methods that maintain the presence of mixing water in the concrete during the early hardening period. These include ponding or immersion, spraying or fogging, and saturated wet coverings. These methods afford some cooling through evaporation, which is beneficial in hot weather.

2. Methods that reduce the loss of mixing water from the surface of the concrete. This can be done by covering the concrete with impervious paper or plastic sheets, or by applying membrane-forming curing compounds.

3. Methods that accelerate strength gain by supplying heat and additional moisture to the concrete. This is usually accomplished with live steam, heating coils, or electrically heated forms or pads.

   The method or combination of methods chosen depends on factors such as availability of curing materials, size, shape, and age of concrete, production facilities (in place or in a plant), esthetic appearance, and economics. As a result, curing often involves a series of procedures used at a particular time as the concrete ages. For example, fog spraying or plastic covered wet burlap can precede application of a curing compound. The timing of each procedure depends on the degree of hardening of the concrete needed to prevent the particular procedure from damaging the concrete surface.

Ponding and Immersion

On flat surfaces, such as pavements and floors, concrete can be cured by ponding. Earth or sand dikes around the perimeter of the concrete surface can retain a pond of water. Ponding is an ideal method for preventing loss of moisture from the concrete; it is also effective for maintaining a uniform temperature in the concrete. The curing water should not be more than about 11 °C (20 °F) cooler than the concrete to prevent thermal stresses that could result in cracking. Since ponding requires considerable labor and supervision, the method is generally used only for small jobs.

The most thorough method of curing with water consists of total immersion of the finished concrete element. This method is commonly used in the laboratory for curing concrete test specimens. Where appearance of the concrete is important, the water used for curing by ponding or immersion must be free of substances that will stain or discolor the concrete. The material used for dikes may also discolor the concrete.

Fogging and Sprinkling

Fogging and sprinkling with water are excellent methods of curing when the ambient temperature is well above freezing and the humidity is low. A fine fog mist is frequently applied through a system of nozzles or sprayers to raise the relative humidity of the air over flatwork, thus slowing evaporation from the surface. Fogging is applied to minimize plastic shrinkage cracking until finishing operations are complete. Once the concrete has set sufficiently to prevent water erosion, ordinary lawn sprinklers are effective if good coverage is provided and water runoff is of no concern. Soaker hoses are useful on surfaces that are vertical or nearly so.

The cost of sprinkling may be a disadvantage. The method requires an ample water supply and careful supervision. If sprinkling is done at intervals, the concrete must be prevented from drying between applications of water by using burlap or similar materials; otherwise alternate cycles of wetting and drying can cause surface crazing or cracking.

Fogging minimizes moisture loss during and after placing and finishing of concrete.

Wet Coverings

Fabric coverings saturated with water, such as burlap, cotton mats, rugs, or other moisture-retaining fabrics, are commonly used for curing . Treated burlaps that reflect light and are resistant to rot and fire are available. The requirements for burlap are described in the Specification for Burlap Cloths Made from Jute or Kenaf, and those for white burlap-polyethylene sheeting are described in ASTM C 171.

Lawn sprinklers saturating burlap with water keep the concrete continuously moist. Intermittent sprinkling is acceptable if no drying of the concrete surface occurs.

Burlap must be free of any substance that is harmful to concrete or causes discoloration. New burlap should be thoroughly rinsed in water to remove soluble substances and to make the burlap more absorbent.

Wet, moisture-retaining fabric coverings should be placed as soon as the concrete has hardened sufficiently to prevent surface damage. During the waiting period other curing methods are used, such as fogging or the use of membrane forming finishing aids. Care should be taken to cover the entire surface with wet fabric, including the edges of slabs. The coverings should be kept continuously moist so that a film of water remains on the concrete surface throughout the curing period. Use of polyethylene film over wet burlap is a good practice; it will eliminate the need for continuous watering of the covering. Periodically rewetting the fabric under the plastic before it dries out should be sufficient. Alternate cycles of wetting and drying during the early curing period may cause crazing of the surface.

Wet coverings of earth, sand, or sawdust are effective for curing and are often useful on small jobs. Sawdust from most woods is acceptable, but oak and other woods that contain tannic acid should not be used since deterioration of the concrete may occur. A layer about 50 mm (2 in.) thick should be evenly distributed over the previously moistened surface of the concrete and kept continuously wet.

Wet hay or straw can be used to cure flat surfaces. If used, it should be placed in a layer at least 150 mm (6 in.) thick and held down with wire screen, burlap, or tarpaulins to prevent its being blown off by wind.

A major disadvantage of moist earth, sand, sawdust, hay, or straw coverings is the possibility of discoloring the concrete.

Impervious Paper

Impervious paper for curing concrete consists of two sheets of kraft paper cemented together by a bituminous adhesive with fiber reinforcement. Such paper, conforming to ASTM C 171, is an efficient means of curing horizontal surfaces and structural concrete of relatively simple shapes. An important advantage of this method is that periodic additions of water are not required. Curing with impervious paper enhances the hydration of cement by preventing loss of moisture from the concrete.

As soon as the concrete has hardened sufficiently to prevent surface damage, it should be thoroughly wetted and the widest paper available applied. Edges of adjacent sheets should be overlapped about 150 mm (6 in.) and tightly sealed with sand, wood planks, pressure-sensitive tape, mastic, or glue. The sheets must be weighted to maintain close contact with the concrete surface during the entire curing period.

Impervious paper can be reused if it effectively retains moisture. Tears and holes can easily be repaired with curing-paper patches. When the condition of the paper is questionable, additional use can be obtained by using it in double thickness.

In addition to curing, impervious paper provides some protection to the concrete against damage from subsequent construction activity as well as protection from the direct sun. It should be light in color and nonstaining to the concrete. Paper with a white upper surface is preferable for curing exterior concrete during hot weather.

Plastic Sheets

Plastic sheet materials, such as polyethylene film, can be used to cure concrete Polyethylene film is a lightweight, effective moisture retarder and is easily applied to complex as well as simple shapes. Its application is the same as described for impervious paper.

Curing with polyethylene film (or impervious paper) can cause patchy discoloration, especially if the concrete contains calcium chloride and has been finished by hardsteel troweling. This discoloration is more pronounced when the film becomes wrinkled, but it is difficult and time consuming on a large project to place sheet materials without wrinkles. Flooding the surface under the covering may prevent discoloration, but other means of curing should be used when uniform color is important.

Polyethylene film should conform to ASTM C 171, which specifies a 0.10-mm (4-mil) thickness for curing concrete, but lists only clear and white opaque film. However, black film is available and is satisfactory under some conditions. White film should be used for curing exterior concrete during hot weather to reflect the sun's rays. Black film can be used during cool weather or for interior locations. Clear film has little effect on heat absorption.

Polyethylene film is an effective moisture barrier for curing concrete and easily applied to complex as well as simple shapes. To minimize discoloration, the film should be kept as flat as possible on the concrete surface.

ASTM C 171 also includes a sheet material consisting of burlap impregnated on one side with white opaque polyethylene film. Combinations of polyethylene film bonded to an absorbent fabric such as burlap help retain moisture on the concrete surface.

Polyethylene film may also be placed over wet burlap or other wet covering materials to retain the water in the wet covering material. This procedure eliminates the labor-intensive need for continuous watering of wet covering materials.

Membrane-Forming Curing Compounds

Liquid membrane-forming compounds consisting of waxes, resins, chlorinated rubber, and other materials can be used to retard or reduce evaporation of moisture from concrete. They are the most practical and most widely used method for curing not only freshly placed concrete but also for extending curing of concrete after removal of forms or after initial moist curing. However, the most effective methods of curing concrete are wet coverings or water spraying that keeps the concrete continually damp. Curing compounds should be able to maintain the relative humidity of the concrete surface above 80% for seven days to sustain cement hydration.

Membrane-forming curing compounds are of two general types: clear, or translucent; and white pigmented. Clear or translucent compounds may contain a fugitive dye that makes it easier to check visually for complete coverage of the concrete surface when the compound is applied. The dye fades away soon after application. On hot, sunny days, use of white-pigmented compounds are recommended; they reduce solar-heat gain, thus reducing the concrete temperature. Pigmented compounds should be kept agitated in the container to prevent pigment from settling out.

Curing compounds should be applied by hand-operated or power-driven spray equipment immediately after final finishing of the concrete. The concrete surface should be damp when the coating is applied. On dry, windy days, or during periods when adverse weather conditions could result in plastic shrinkage cracking, application of a curing compound immediately after final finishing and before all free water on the surface has evaporated will help prevent the formation of cracks. Power-driven spray equipment is recommended for uniform application of curing compounds on large paving projects. Spray nozzles and windshields on such equipment should be arranged to prevent wind-blown loss of curing compound.

Normally only one smooth, even coat is applied at a typical rate of 3 to 4 m2 per liter (150 to 200 sq ft per gallon); but products may vary, so manufacturer's recommended application rates should be followed. If two coats are necessary to ensure complete coverage, for effective protection the second coat should be applied at right angles to the first. Complete coverage of the surface must be attained because even small pinholes in the membrane will increase the evaporation of moisture from the concrete.

Curing compounds might prevent bonding between hardened concrete and a freshly placed concrete overlay. And, most curing compounds are not compatible with adhesives used with floor covering materials. Consequently, they should either be tested for compatibility, or not used when bonding of overlying materials is necessary. For example, a curing compound should not be applied to the base slab of a two-course floor. Similarly, some curing compounds may affect the adhesion of paint to concrete floors. Curing compound manufacturers should be consulted to determine if their product is suitable for the intended application.

Fig: Liquid membrane-forming curing compounds should be applied with uniform and adequate coverage over the entire surface and edges for effective, extended curing of concrete.

Curing compounds should be uniform and easy to maintain in a thoroughly mixed solution. They should not sag, run off peaks, or collect in grooves. They should form a tough film to withstand early construction traffic without damage, be nonyellowing, and have good moisture-retention properties.

Caution is necessary when using curing compounds containing solvents of high volatility in confined spaces or near sensitive occupied spaces such as hospitals because evaporating volatiles may cause respiratory problems. Applicable local environmental laws concerning volatile organic compound (VOC) emissions should be followed.

Curing compounds should conform to ASTM C 309. A method for determining the efficiency of curing compounds, waterproof paper, and plastic sheets is described in ASTM C 156. ASTM C 1151, discontinued in 2000, also evaluates the effectiveness of curing compounds. Curing compounds with sealing properties are specified under ASTM C 1315.

Internal Moist Curing

Internal moist curing refers to methods of providing moisture from within the concrete as opposed to outside the concrete. This water should not effect the initial water to cement ratio of the fresh concrete. Lightweight (low-density) fine aggregate or absorbent polymer particles with an ability to retain a significant amount of water may provide additional moisture for concretes prone to self desiccation. When more complete hydration is needed for concretes with low water to cement ratios (around 0.30 or less), 60 kg/m^3 to 180 kg/m^3 (100 lb/yd^3 to 300 lb/yd^3) of saturated lightweight fine aggregate can provide additional moisture to extend hydration, resulting in increased strength and durability. All of the fine aggregate in a mixture can be replaced with saturated lightweight fine aggregate to maximize internal moist curing. Internal moist curing must be accompanied by external curing methods.

Forms Left in Place

Forms provide satisfactory protection against loss of moisture if the top exposed concrete surfaces are kept wet. A soaker hose is excellent for this. The forms should be left on the concrete as long as practical.

Wood forms left in place should be kept moist by sprinkling, especially during hot, dry weather. If this cannot be done, they should be removed as soon as practical and another curing method started without delay. Color variations may occur from formwork and uneven water curing of walls.

Steam Curing

Steam curing is advantageous where early strength gain in concrete is important or where additional heat is required to accomplish hydration, as in cold weather.

Two methods of steam curing are used: live steam at atmospheric pressure (for enclosed cast-in-place structures and large precast concrete units) and high-pressure steam in autoclaves (for small manufactured units). Only live steam at atmospheric pressure will be discussed here.

A typical steam-curing cycle consists of (1) an initial delay prior to steaming, (2) a period for increasing the temperature, (3) a period for holding the maximum tem- perature constant, and (4) a period for decreasing the temperature. A typical atmospheric steam-curing cycle is shown in figure.

Steam curing at atmospheric pressure is generally done in an enclosure to minimize moisture and heat losses. Tarpaulins are frequently used to form the enclosure. Application of steam to the enclosure should be delayed until initial set occurs or delayed at least 3 hours after final placement of concrete to allow for some hardening of the concrete. However, a 3- to 5- hour delay period prior to steaming will achieve maximum early strength.

Steam temperature in the enclosure should be kept at about 60 °C (140 °F) until the desired concrete strength has developed. Strength will not increase significantly if the maximum steam temperature is raised from 60 °C to 70°C (140°F to 160°F). Steam-curing temperatures above 70 °C (160 °F) should be avoided; they are uneconomical and may result in damage. It is recommended that the internal temperature of concrete not exceed 70 °C (160 °F) to avoid heat induced delayed expansion and undue reduction in ultimate strength. Use of concrete temperatures above 70 °C (160 °F) should be demonstrated to be safe by test or historic field data.

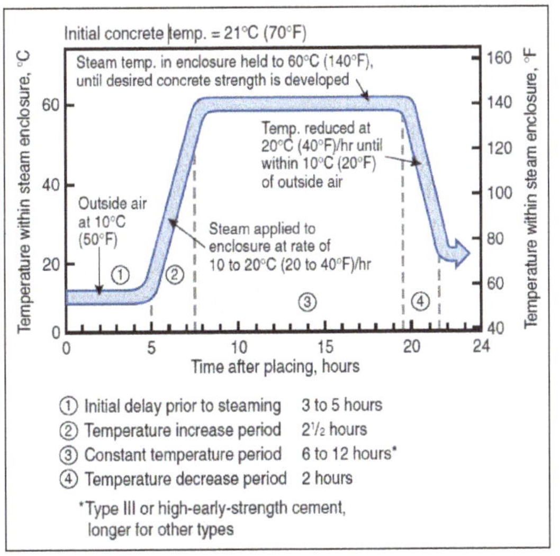

A typical atmospheric steam-curing cycle.

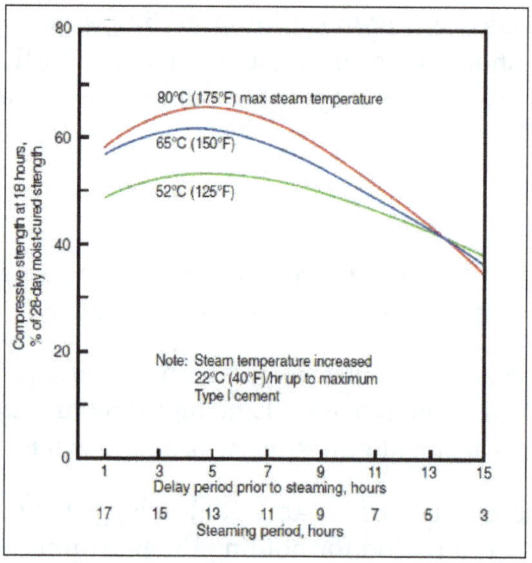

Relationship between strength at 18 hours and delay period prior to steaming. In each case, the delay period plus the steaming period totaled 18 hours.

Concrete temperatures are commonly monitored at the exposed ends of the concrete element. Monitoring air temperatures alone is not sufficient because the heat of hydration may cause the internal temperature of the concrete to exceed 70 °C (160 °F). Besides early strength gain, there are other advantages of curing concrete at temperatures of around 60 °C (140 °F); for example, there is reduced drying shrinkage and creep as compared to concrete cured at 23 °C (73°F) for 28 days.

Excessive rates of heating and cooling should be avoided to prevent damaging volume changes. Temperatures in the enclosure surrounding the concrete should not be increased or decreased more than 22 °C to 33 °C (40 °F to 60 °F) per hour depending on the size and shape of the concrete element.

The curing temperature in the enclosure should be held until the concrete has reached the desired strength. The time required will depend on the concrete mixture and steam temperature in the enclosure.

Insulating Blankets or Covers

Layers of dry, porous material such as straw or hay can be used to provide insulation against freezing of concrete when temperatures fall below 0°C (32 °F).

Formwork can be economically insulated with commercial blanket or batt insulation that has a tough moistureproof covering. Suitable insulating blankets are manufactured of fiberglass, sponge rubber, cellulose fibers, mineral wool, vinyl foam, and open-cell poly-urethane foam. When insulated formwork is used, care should be taken to ensure that concrete temperatures do not become excessive.

Framed enclosures of canvas tarpaulins, reinforced polyethylene film, or other materials can be placed around the structure and heated by space heaters or steam. Portable hydronic heaters are used to thaw subgrades as well as heat concrete without the use of an enclosure.

Electrical, Oil, Microwave and Infrared Curing

Electrical, hot oil, microwave and infrared curing methods have been available for accelerated and normal curing of concrete for many years. Electrical curing methods include a variety of techniques: (1) use of the concrete itself as the electrical conductor, (2) use of reinforcing steel as the heating element, (3) use of a special wire as the heating element, (4) electric blankets, and (5) the use of electrically heated steel forms (presently the most popular method). Electrical heating is especially useful in cold- weather concreting. Hot oil may be circulated through steel forms to heat the concrete. Infrared rays and microwave have had limited use in accelerated curing of concrete. Concrete that is cured by infrared methods is usually under a covering or enclosed in steel forms. Electrical, oil, and infrared curing methods are used primarily in the precast concrete industry.

Curing Period and Temperature

The period of time that concrete should be protected from freezing, abnormally high temperatures, and against loss of moisture depends upon a number of factors: the type of cementing materials used; mixture proportions; required strength, size and shape of the concrete member; ambient weather; and future exposure conditions. The curing period may be 3 weeks or longer for lean concrete mixtures used in massive structures such as dams; conversely, it may be only a few days for rich mixes, especially if Type III or HE cement is used. Steam-curing periods are normally much shorter, ranging from a few hours to 3 days; but generally 24-hour cycles are used. Since all the desirable properties of concrete are improved by curing, the curing period should be as long as necessary.

For concrete slabs on ground (floors, pavements, canal linings, parking lots, driveways, sidewalks) and for structural concrete (cast-in-place walls, columns, slabs, beams, small footings, piers, retaining walls, bridge decks), the length of the curing period for ambient temperatures above 5°C (40°F) should be a minimum of 7 days; additional time may be needed to attain 70% of the specified compressive or flexural strength. When the daily mean ambient temperature is 5°C (40°F) or lower, ACI Committee 306 recommendations for curing should be followed to prevent damage by freezing.

A higher curing temperature provides earlier strength gain in concrete than a lower temperature but it may decrease 28-day strength as shown in figure. If strength tests are made to establish the time when curing can cease or forms can be removed, representative concrete test cylinders or beams should be fabricated in the field, kept adjacent to the structure or pavement they represent, and cured by the same methods. Equipment is available that can monitor internal concrete temperatures and match that temperature in the concrete cylinder curing box; this is the most accurate means of representing in-place concrete strengths. Cores, cast-in-place removable cylinders, and nondestructive testing methods may also be used to determine the strength of a concrete member.

Since the rate of hydration is influenced by cement type and the presence of supplementary cementing materials, the curing period should be prolonged for concretes made with cementing materials possessing slow-strength- gain characteristics. For mass concrete (large piers, locks, abutments, dams, heavy footings, and massive columns and transfer girders) in which no pozzolan is used as part of the cementitious material, curing of unreinforced sections should continue for at least 2 weeks. If the mass concrete contains a pozzolan, minimum curing time for unreinforced sections should be extended to 3 weeks. Heavily-reinforced mass concrete sections should be cured for a minimum of 7 days.

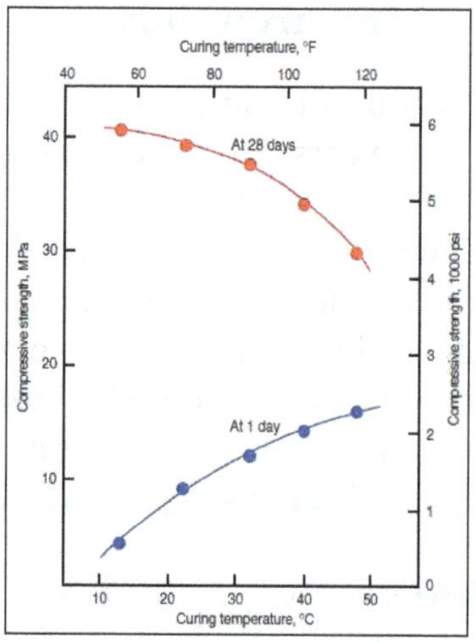

One-day strength increases with increasing curing temperature, but 28-day strength decreases with increasing curing temperature.

During cold weather, additional heat is often required to maintain favorable curing temperatures of 10 °C to 20 °C (50 °F to 70 °F). Vented gas or oil-fired heaters, heating coils, portable hydronic heaters, or live steam can be used to supply the required heat. In all cases, care must be taken to avoid loss of moisture from the concrete. Exposure of fresh concrete to heater or engine exhaust gases must be avoided as this can result in surface deterioration and dusting (rapid carbonation).

High-early-strength concrete can be used in cold weather to speed-up setting tim e and strength development. This can reduce the curing period, but a minimum temperature of 10 °C (50 °F) must be maintained.

For adequate deicer scale resistance of concrete, the minimum curing period generally corresponds to the time required to develop the design strength of the concrete at the surface. A period of air-drying after curing will enhance resistance to scaling. This drying period should be at least 1 month of relatively dry weather before the application of deicing salts.

Sealing Compounds

Sealing compounds (sealers) are liquids applied to the surface of hardened concrete to reduce the penetration of liquids or gases such as water, deicing solutions, and carbon dioxide that cause freeze-thaw damage, corrosion of reinforcing steel, and acid attack. In addition, sealing compounds used on interior floor slabs reduce dusting and the absorption of spills while making the surface easier to clean.

Sealing compounds differ in purpose from curing compounds; they should not be confused as being the same. The primary purpose of a curing compound is to retard the loss of water from newly placed concrete and it is applied immediately after finishing. Surface sealing compounds on the other hand retard the penetration of harmful substances into hardened concrete and are typically not applied until the concrete is 28 days old. Surface sealers are generally classified as either film-forming or penetrating.

Sealing exterior concrete is an optional procedure generally performed to help protect concrete from freezethaw damage and chloride penetration from deicers. Curing is not optional when using a sealer; curing is necessary to produce properties needed for concrete to perform adequately for its intended purpose. Satisfactory performance of exterior concrete still primarily depends on an adequate air void system, sufficient strength, and the use of proper placing, finishing and curing techniques. However, not all concrete placed meets those criteria; surface sealers can help improve the durability of these concretes.

Film-forming sealing compounds remain mostly on the surface with only a slight amount of the material penetrating the concrete. The relatively large molecular structure of these compounds limits their ability to penetrate the surface. Thinning them with solvents will not improve their penetrating capability. These materials not only reduce the penetration of water, they also protect against mild chemicals; furthermore, they prevent the absorption of grease and oil as well as reduce dusting under pedestrian traffic.

Surface sealers consist of acrylic resins, chlorinated rubber, urethanes, epoxies, and alpha methyl styrene. The effectiveness of film-forming sealers depends on the continuity of the layer formed. Abrasive grit and heavy traffic can damage the layer requiring the reapplication of the material. Consult manufacturers' application recommendations because some of these materials are intended for interior use only and may yellow and deteriorate under exposure to ultraviolet light.

The penetrating sealer that has been used most extensively for many years is a mixture of 50 percent boiled linseed oil and 50 percent mineral spirits. Although this mixture is an effective sealer, it has two main disadvantages: it darkens the concrete, and periodic reapplication is necessary for long-term protection.

A new generation of water-repellent penetrating sealers have a very small molecular size that allows penetration and saturation of the concrete as deep as 3 mm (1/8 in.). The two most common are silane and siloxane, compounds which are derived from the silicone family. These sealers allow the concrete to breath, thus preventing a buildup of vapor pressure between the concrete and sealer that can occur with some film-forming materials. Because the sealer is embedded within the concrete, making it more durable to abrasive forces or ultraviolet deterioration, it can provide longer lasting protection than film-forming sealers. However, periodic retreatment is recommended. In northern states and coastal areas silanes and siloxanes are popular for protecting bridge decks and other exterior structures from corrosion of reinforcing steel caused by chloride infiltration from deicing chemicals or sea spray.

Application of any sealer should only be done on concrete that is clean and allowed to dry for at least 24 hours at temperatures above 16 °C (60 °F). At least 28 days should be allowed to elapse before applying sealers to new concrete. Penetrating sealers cannot fill surface voids if they are filled with water. Some surface preparation may be necessary if the concrete is old and dirty. Concrete placed in the late fall should not be sealed until spring because the sealer may cause the concrete to retain water that may exacerbate freeze-thaw damage.

Sustainability in Concrete Construction

The sustainability of the cement and concrete industries is imperative to the well-being of our planet and to human development. However, the production of portland cement, an essential constituent of concrete, leads to the release of a significant amount of CO_2 and other greenhouse gases (GHGs). The environmental issues associated with CO_2 will play a leading role in the sustainable development of the cement and concrete industry during this century. One of the biggest threats to the sustainability of the cement industry is the dwindling amount of limestone in some geographical regions. Limestone is essential to the production of Portland cement. As limestone becomes a limited resource, employment and construction associated with the concrete industry will decline. Therefore, those involved with these industries must develop new techniques for creating concrete with minimal use of limestone. Not only is concrete production a valuable source of societal development, but it is also a significant source of employment. Concrete is the world's most consumed man-made material. Therefore, not only to create sustainable societal development but also to sustain employment—such as batch plant operators, truck drivers, ironworkers, labourers, carpenters, finishers, equipment operators, and testing technicians, as well as professional engineers, architects, surveyors, and inspectors—the concrete industry must continue to evolve with the changing needs and expectations of society worldwide.

Limestone is used to manufacture portland cement and it is also used as aggregate in concrete. Currently, portland cement is the most commonly used material in producing concrete. Some geographical regions are running out of limestone resources to produce cement. And major metropolitan areas are running out of materials to use as aggregates for making concrete. Sustainability requires those in the construction industry to consider the entire life cycle—including construction, maintenance, demolition, and recycling of buildings.

A sustainable concrete structure is one that is constructed so that the total societal impact during its entire life cycle is minimal. Designing with sustainability in mind includes accounting for the short-term and long-term consequences of the structure. To decrease the long-term impact of structures, the creation of durable structures is paramount.

Building in a sustainable manner and scheduling appropriate building maintenance are significant in the "new construction ideology" of this new century. In particular, to build in a sustainable manner means to focus attention on the effects on human health, energy conservation, and physical, environmental, and technological resources for new and existing buildings. It is also important to take into account the impact of construction technologies and methods when creating sustainable structures. An integrated sustainable design process can reduce project costs and operating costs of the building or the infrastructure construction.

Many challenges are associated with portland cement production. Of these, energy and resource conservation, the cost of producing portland cement, and GHG emissions are the most significant. Therefore, supplementary cementing materials such as fly ash and ground granulated blast-furnace slag should replace larger amounts of portland cement in concrete. However, before any construction occurs, all aspects of the building materials to be used should be evaluated. To build structures and infrastructures that are cost-efficient, environmentally friendly, and durable, the impact of the building materials on local and worldwide air conditions must be examined. At the current rate of increase of cement production, worldwide cement production is expected to rise from about 2.5 billion t in 2006 to about 5 billion t by 2020. Thus, CO2 emission caused by portland cement production is expected to rise by 100% from the current level by 2020.

For each metric ton of Portland cement clinker, 1.5 to 10 kg of NOx is released into the atmosphere. In 2000, worldwide cement clinker production was approximately 1.5 billion t. In 2000, between 2.3 and 15 million t of NOx were produced to make portland cement clinker. Clinker production in the United States was about 85 million t in 2005; and therefore, about 125 to 850 thousand t of NOx were produced in the United States to make portland cement. If the challenges associated with reducing CO_2, NOx, and other GHGs are to be met, then the concrete industry must develop other materials to replace portland cement. The use of blended cements and organic chemical admixtures must be significantly increased for sustainability of the cement and concrete industries.

Portland Cement

Portland cement is not an environmentally friendly material, because its manufacture creates greenhouse gas emissions; it also reduces the supply of good-quality limestone and clay. As good engineers, we must reduce the use of Portland cement in concrete. We must use more blended cements to reduce the need for Portland cement clinker per metric ton of blended cement produced by blending with the clinker other PMs, such as coal or wood fly ash, slag, silica fume, and such other PMs as finely ground waste glass. As a cement production feed material, instead of clay, industrial by-products such as used foundry sand or coal combustion products (CCPs) such as fly ash should be used in the optimum possible quantity.

The most energy-intensive stage of portland cement production is during clinker production. It accounts for all but about 10% of the energy use and nearly all of the GHGs produced by cement production. Kiln systems evaporate inherent water from the raw meal and calcine the carbonate constituents during clinker pre-processing. Sources of CO_2 and GHG emissions in the manufacturing of portland cement are as follows:

- From calcinations of limestone = 50–55%.
- From fuel combustion = 40–50%.
- From use of electric power = 0–10%.

Innovative Cement Products

Although the embodied energy linked to concrete production is low, PMs, especially coal fly ash, have been used by the concrete industry for more than 70 years. Their use can contribute to a further reduction of concrete's embodied energy. When used wisely and judiciously, PMs can improve the long-term properties of concrete. Fly ash can and does regularly replace portland cement in concrete.

One process that is even more environmentally friendly and productive is the use of blended cements. Blended cements have been used for many decades. Blended cements are made when various amounts of clinker are blended or interground with one or more additives, including fly ash, natural pozzolonas, slag, silica fume, and other PMs. Blended cements allow for a reduction in the energy used and reduce GHG emissions.

Most innovative concrete mixtures make use of PMs to partially replace cement. The advantages of blended cements include increased production capacity, reduced GHG emissions, reduced fuel consumption in the final cement production, and recycling of PMs. The manufacture of portland cement is the third most energy intensive process, after aluminium and steel manufacture. In fact, for each metric ton of Portland cement, about 5-1/2 million BTU of energy are needed .

Although cement production is energy-inefficient, major initiatives have reduced energy consumption. Of these, the most significant has been the replacement of wet production facilities with dry processing plants. In addition, the cement industry has also moved away from petroleum-based fuel use.

Despite these advances, some shortcomings still exist when energy use is evaluated for the concrete industry. Dry process cement plants use preheaters, which increase the alkali content of cement. Thus, cement producers need to continue to develop ways to control the alkali content without increasing the energy consumption levels of the cement. Furthermore, current innovations and energy savings are linked to the amount of energy consumption by converting wet-process to dry-process cement production and the number of preheaters needed to complete the process. For each million metric ton of capacity, a new Portland cement plant costs more than $225 million. The cost associated with producing portland cement, along with the CO_2 emissions and energy issues, makes it unlikely that developing countries will be able to employ modern technology to reduce GHGs. Also, government regulations of GHGs will likely force the cement industry to create blended cements and use supplementary materials for blended cements to meet societal development needs.

To produce 1 t of portland cement, 1.6 t of raw materials are needed. These materials include good-quality limestone and clay. Therefore, to manufacture the current worldwide production of 2.5 billion t of cement annually, at least 4 billion t of raw materials are needed. As good engineers, we must employ environmentally friendly materials to reduce the use of portland cement by replacing a major part of portland cement with PMs for use in concrete. In the United States, such materials—primarily fly ash, slag, silica fume, natural pozzolonas, rice-husk ash, wood ash, and agricultural- products ash—are available for up to 70% replacement. All these materials can be used to supplement the use of cement in concrete mixtures while improving the concrete product durability. One of the important benefits of the increased use of other types of cementitious materials !such as PMs" is the reduction of GHG emissions. With replacement of cement with other recyclable resources, worldwide CO_2 emissions would be reduced. A replacement of 50% of cement worldwide by other cementitious materials would reduce CO_2 emissions by more than 1 billion t. This is equivalent to removing approximately one-quarter of all automobiles in the world . Fly ash production in the United States in 2005 was estimated at 64 million t by the American Coal Ash Association; by 2010 it is estimated to be 70 million t. Portland cement use in the United States in 2005 was estimated at more than 125 million t; by 2010 it is estimated to be 140 million t. The fly ash disposal challenge and the limited availability of Portland cement have the same solution: replace large amounts of Portland cement with fly ash to create durable and sustainable concrete.

Waste Materials

Engineers, architects, planners, and builders should reuse industrial by-products and postconsumer wastes in concrete and other cement-based construction materials. Postconsumer wastes that should be considered for use in concrete include glass, plastics, tires, demolished concrete, and clay bricks. To do this successfully, designers must watch for harmful hydration reactions of portland cement and changes in the volume of concrete. The recycling of industrial by-products has been well established in the cement and concrete industries over the past several decades. The use of coal fly ash in concrete began in the 1930s, but volcanic ash has been used in mortar and primitive concrete for several millennia in Egypt, Italy, Mexico, and India. The use of by-products such as rice-husk ash, wood ash, silica fume, and other pozzolonic materials, in addition to coal fly ash, can help reduce the need for portland cement in addition to creating more durable concrete and reducing greenhouse gas emissions. This will also contribute to the improvement of air quality, reduction of solid Wastes, and sustainability of the cement and concrete industry.

In summary, for sustainability of the cement and concrete industries, use less portland cement; use less water; use application-specific high-quality, durable aggregates; and use organic chemical admixtures. Fundamental laws of nature state that we cannot create or destroy matter; we can only affect how it is organized, transformed, and used. To manage natural resources, humanity must obey the rules of nature: use only what you need and never use a resource faster than nature can replenish it. Resources are extracted from the earth at a rate of 20% more than the earth produces. Therefore, what is consumed in 12 months will take 14.4 months to be replenished. The use of sustainable development procedures will reduce that rate. The issue is not environment vs. development or ecology vs. economy; the two can be (and must be) integrated.

Shape Optimization – Lightening of RC Structure

Shape optimization can result in subtle lightened cross-sections. Their lower weight imposes lower load on supporting structures. Application of high performance concrete enables additional savings due to higher reduction of cross section dimensions.

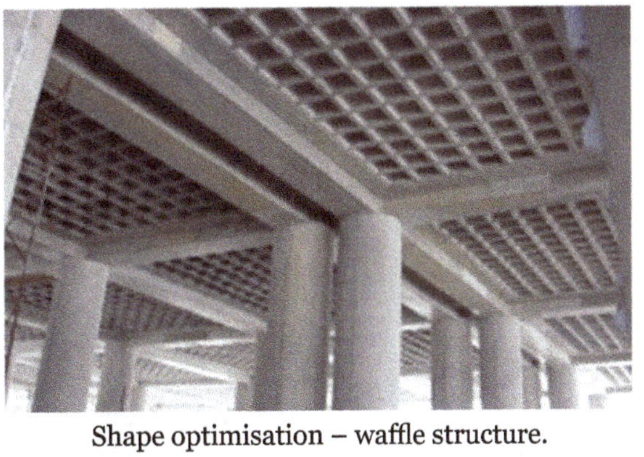

Shape optimisation – waffle structure.

Installation elements from recycled plastic.

Cross section shape can be created using moulds, various types of lightening elements or by application of light-weight concrete. Mentioned techniques can lead to material savings from 30 to 60% .

Optimization of Concrete Mixture – High Performance Concrete

Utilizing new composite materials with significantly better physical characteristics creates realistic assumptions of achieving substantial effect from the perspective of material and energy savings. Some examples from abroad show that high performance concrete can be used for optimized shapes of RC (reinforced concrete) elements, which can be very subtle (wall thickness of 30 mm and less) due to their mechanical parameters. Nowadays, concretes with compressive strength around 100 MPa are used in some cases, also UHPC (UHPC – Ultra High Performance Concrete) with compressive strength over 150 MPa have already been used for some special application (e.g. Ductal, Lafarge, France). These kinds of materials enable design with reduced material consumption and thus with lower environmental impacts.

Utilization of Recycled Concrete and other Secondary Materials

Some secondary materials (waste recycling products from other processes) can be used for cement and concrete production, for production of mould elements or other components. Fly ash, slag or microsilica are used to increase strength and workability. They decrease consumption of energy intensive Portland cement due to their cementitious properties. Recycled concrete from demolition can be used as backfill replacing natural aggregate. Recycled aggregate can be used to some extent as substitute to natural aggregate in new concrete.

Concrete for Energy Effective Buildings

Accumulative properties of RC structures can significantly contribute to thermal stability improvement of buildings indoor environment. This will mainly affect buildings that need cooling in summer.

Subtle RC frame for residential buildings – used for family house in low energy standard.

With higher demands on thermal insulation parameters of building envelopes increases also their thickness. The envelope thickness represents restricting parameter of developer plan in regions with regulated size of built-up area. In these cases, subtle RC structures become a great advantage in the form of slender load-bearing wall or subtle RC frame.

Precast Concrete Structures with Integrated Functions

Prefabrication enables design of integrated structural elements of more complicated shapes adjusted to more effective utilization of concrete structure also for other functions. In a case of hollow core concrete panels there can be utilized the accumulative potential of surface layer in hollows. Panels can contain built-in distribution system or tubes for cable protection for subsequent installation of electricity, water, ventilation and heating distribution systems.

Timber-Concrete Composite Structures

Combination of concrete and timber elements can lead to advantageous structural static and environmental solutions. Especially in the case of multi-storey timber buildings there can concrete floor structure or its upper deck section significantly improve acoustic and fire safety parameters and at the same time it can ensure horizontally rigid floor structure from the perspective of spatial rigidity.

Textile Reinforced Concrete Utilization

Textile reinforced concrete (TRC) has been developed for very thin plate or shell elements (thickness 12 to 25 mm) where the use of steel reinforcement is not applicable because of a risk of corrosion due to a thin concrete cover layer. The concept of using TRC was introduced in order to reduce the thickness and thus weight of elements and consequently consumption of concrete and associated environmental impacts. This concept is mostly used for shells with particular shapes, for strengthening of structural elements or for thin façade panels. Figure shows testing of TRC plate from UHPC and two layers of 2D textile reinforcement in Experimental Centre at CTU in Prague. On the Figure there is a prototype of façade panel from white TRC (13 mm thick TRC plate is fixed to steel bearing frame).

Testing of TRC thin plate element.

Prototype of façade panel from white TRC.

Quality Assurance for Concrete Construction

Quality of concrete construction on site can be accomplished in three distinct stages as follow:

1. Quality control before concreting.
2. Quality control during concreting.
3. Quality control after construction.

Stage-1 (Quality Control before Concreting)

This stage of quality control consists of two steps:

- Checking of specification requirements regarding excavation, forms, reinforcement and embedded fixtures etc.
- Control test on concrete ingredients (i.e. on cement, aggregate & water).

Cement

Quality of cement is ascertained by making compressive strength tests on cement cubes. However for effective control cement:

- Should be tested initially once for each source and subsequently once for every two months.
- Should be protected from moisture.
- Should be retested after 3 months of storage, if long storage in unavoidable.
- Should be rejected if large lump are found in cement bags.

Aggregate

Concrete aggregates should confirm to specified values as per standard specification.

The quality of concrete is affected by different physical and mechanical properties of aggregate, i.e. shape, grading, durability, specific gravity and water absorption etc. these properties of aggregated should be tested before using it for concrete production.

The quantity of deleterious materials and organic impurities should also be tested.

Bulking of sand is also an important property in several ways. It gives wrong results when volume batching is done. It increases water cement ratio which in turn reduces strength.

For effective control aggregates:

- Are required to be tested once initially for approval of source.
- Should subsequently be tested once or twice daily for moisture content and allowance should be made for moisture content of aggregates.

Water

The quality of water should be checked for the requirements as specified in respective standard. Chemical analysis shall be conducted for approval of source. In case of suspended impurities, it is necessary to store water for some time to allow them to settle down. In case of doubt concrete cubes made with this water are tested. Average 28 days compressive strength of at least three cubes or cylinders or specified size, prepared with water proposed to be used shall not be less than 90% of the average strength of three similar concrete cubes prepared with distilled water.

Stage-2 (Quality Control during Concreting)

Careful supervision during concrete manufacture is necessary for all concreting operations such as batching, mixing, transporting, laying, compacting and curing. Following precautions should be taken during concreting operation:

1. The concrete mix should be designed in the laboratory with the materials to be used on site.
2. As far as possible concrete should be batched by weight. If weight batching is not possible, then volume batching may be permitted through proper supervision in the presence of engineer in charge.
3. During mixing the mixer should be charged to its full capacity. The materials should be fed in proper sequence. The speed of the mixer should be range from 15 to 20 revolutions per minute. The mixing time should not be less than 2 minutes in any case. Segregation should be avoided while unloading the concrete from the mixer.
4. Workability of concrete is an important property of concrete while concrete is in its fresh state. Therefore slump test or compaction factor test should be performed to check workability of concrete. About three tests should be carried out for every 25 m^3 of concrete.
5. Care should be taken so that no segregation takes place during transportation of concrete.
6. Concrete should not be dropped from a height of more than 1 m. if the drop height exceeds 1 m chutes should be used.
7. To avoid rehandling of concrete it should be placed at its final position as far as possible.
8. Vibrators should be used for compacting concrete. The insertion spacing of internal vibrators should not be more than 0.6 m. It should be drawn out slowly so that no holes remain in the concrete. The frequency of vibrators should not be less than 7000 cycles/minutes.
9. Curing should be done for a specified period so that concrete develops requisite strength. Concrete should be covered with hessian as soon as it becomes hard.
10. The form work should correspond to final form of the structure. It should be checked before concreting is started. The inside of the forms should be cleaned and oiled. The forms should be removed after the specified period.
11. Concrete should be protected from hot and cold weather at early ages. Concreting should not be done at temperature below 4.5 °C and above 40 °C In very hot weather water and aggregates should be cooled. Retarders of approved quality can be used.
12. In very cold weather water and aggregates should be heated. Accelerators of approved quality can also be used.

Stage-3 (Quality Control after Construction)

Once the concrete is laid and compacted, compression tests are made on the cubes made out of this concrete. For ordinary concrete, cubes are made out of the concrete made at work site.

The hardened concrete has to be checked for trueness in dimensions, shape and sizes as per design specification. General surface appearance of concrete should also be checked. Dimensions are ascertained by different measurements. Reinforcement should have adequate concrete cover and if the reinforcement is visible in part of a structure, the part should be rejected or necessary actions should be taken accordingly.

Concrete strength is normally to be ascertained from cube or cylinder samples tested at 28 days. In case the strength obtained is less than the specific minimum, one or more of following steps may be taken:

- Load test and measurement of deflection and strain (the quality of the structure can then be ascertained by calculating back the concrete strength).
- Cutting cores from the structures and testing them for strength.
- Non destructive tests like Schmidt rebound hammer or ultrasonic pulse velocity test. These tests give only a very rough idea and are primarily used to ascertain the uniformity of construction.
- Chemical analysis of hardened concrete.

References

- Stone: understandconstruction.com, Retrieved 25 July, 2019
- Type-stone-construction: constructionheadline.com, Retrieved 5 January, 2019
- Stone-masonry: civilseek.com, Retrieved 7 April, 2019
- Pros-and-cons-of-building-a-stone-house: edarticle.com, Retrieved 17 July, 2019
- Concrete: madehow.com, Retrieved 28 February, 2019
- Grade-of-concrete, concrete: theconstructor.org, Retrieved 30 August, 2019
- Different-types-of-concrete: civilseek.com, Retrieved 24 January, 2019
- Concrete-frame-structures: understandconstruction.com, Retrieved 1 May, 2019
- Concrete-batching-methods, concrete: theconstructor.org, Retrieved 11 March, 2019
- Different-methods-mixing-concrete: gharpedia.com, Retrieved 15 August, 2019
- Transportation-and-placement-of-concrete-materials-concrete-technology, concrete, concrete-technology: engineeringenotes.com, Retrieved 5 March, 2019
- Quality-control-concrete-construction-work: civilblog.org, Retrieved 4 June, 2019

5

Wood and Steel Frame Construction

Frame construction involves the fitting together of pieces in order to give shape or support to a structure. Primarily wood or steel are used as framing materials. The topics elaborated in this chapter will help in gaining a better perspective about wood and steel frame construction as well as the different properties of wood and steel.

Wood

Wood has been used as a building material for thousands of years, being second only to stone in terms of its rich and storied history in the world of construction. The chemical properties of wood are inherently complex, but even in spite of this challenge, human beings have successfully harnessed the unique characteristics of wood to build a seemingly unlimited variety of structures. This exceptionally versatile material is commonly used to build houses, shelters and boats, but it is also extensively used in the furniture and home decor industry as well.

Perhaps one of the biggest advantages of using wood as a building material is that it is a natural resource, making it readily available and economically feasible. It is remarkably strong in relation to its weight, and it provides good insulation from the cold. Wood is highly machinable, and can be fabricated into all kinds of shapes and sizes to fit practically any construction need. Wood is also the perfect example of an environmentally sustainable product; it is biodegradable and renewable, and carries the lowest carbon footprint of any comparable building material. In addition, no high-energy fossil fuels are required to produce wood, unlike other common building materials such as brick, steel or plastic.

Lumber or Timber

The words "lumber" and "timber" are often used interchangeably to refer to wood used in construction work, but there has been considerable debate as to which term should apply in a given scenario. Pieces of wood that are smaller than 5 inches wide by 5 inches thick (regardless of length) are generally referred to as lumber. These pieces are machine-planed and sawn to fit certain dimensional specifications (e.g., 2 x 4", 2 x 8", etc.) and are primarily used in residential construction. Pieces of wood over 5 inches wide by 5 inches thick (regardless of length) are referred to as timber and any timber pieces that exceed 8" wide by 8" thick are referred to as beams. As timber pieces are larger in dimension, they are often used to construct the frames of large structures such as buildings and bridges. Timber is also commonly utilized in large quantities for railroad ties, mine shaft supports and crossbeams on utility poles.

Another type of wood commonly used in construction is known as engineered wood. As its name implies, engineered wood is the product of a more intricate fabrication process in which various wood strands, fibers, veneers, or other forms of wood are glued together to form a type of composite material that is used for specific construction applications. Common examples of engineered wood include plywood, glued laminated timber (a.k.a. "glulam"), oriented strand board, fiberboard, and particle board. Engineered wood products are commonly used in a wide variety of residential, commercial and industrial construction projects.

Types of Wood

Wood has traditionally been classified into two primary categories: Hardwood (any leaf-bearing tree) and softwood (any cone-bearing tree). As with most other general classifications, this can get somewhat confusing due to the fact that there are some leaf-bearing trees that can have relatively soft wood, while some coniferous trees that can have rather hard wood. Generally speaking, however, hardwoods are by and large considered to be heavier and more dense than softwoods. Hardwoods are commonly used in the construction of walls, ceilings and floors, while softwoods are often used to make doors, furniture and window frames. Some examples of the most popular hardwoods include oak, maple, mahogany, cherry, walnut, and teak. Commonly used softwoods include pine, hickory, beach, ash, birch, and cedar.

Lumber Grades

The National Hardwood Lumber Association (NHLA) of America has created a grading system to rate various types of lumber, primarily based on the amount of defects that can be found in a board. Below is a brief summary of NHLA grades for both hardwood and softwood lumber.

Hardwoods

1. First and Seconds (FAS) - This is the highest grade possible for hardwood lumber, and is mainly suited for high-quality furnishings, solid wood mouldings and interior joinery. Contains 83% usable material on one face (minimum 6" x 8" board size).
2. Select (Sel) - Also contains 83% usable material, but for a smaller minimum board size (4" x 6") than FAS.
3. 1 Common (1 Com) - Contains 66% usable material on a 3" x 4" board face.
4. 2 Common (2 Com) - Contains 50% usable material on a 3" x 4" board face.

Softwoods

1. C Select - Almost completely free of all defects; commonly used for cabinets and interior trim.
2. D Select - Comparable to C Select, but may contain small knots (no bigger than the size of a dime).
3. 1 Common - Contains small, tight knots that won't fall out; offers a high-quality knotty appearance (e.g., pine).
4. 2 Common - Very similar to 1 Common, but with slightly larger knots; often used in shelving and paneling.
5. 3 Common - Larger knots that what are found in 2 Common; typically used for crates, boxes and fences.

Benefits of Wood in Construction

Wood carries several benefits that make it an excellent candidate for use in a wide array of construction projects. One such benefit is its thermal properties, which give it an advantage in terms of its resistance to high temperatures. Unlike steel, which can expand or even collapse in high heat, wood actually dries out and becomes stronger as the heat increases. In addition, the heat conductivity of wood is relatively low in comparison to other materials such as aluminum, marble, steel, or glass. This gives wood an advantage in terms of being used in various applications such as matches, hardware equipment handles, wall coverings, and ceilings.

Wood also contains highly-sought-after acoustic properties. It can absorb sound and echoes, and is a favorite material of choice for the construction of structures where proper acoustics is important, such as concert halls. Wood is resistant to electrical currents, making it an optimal material for electrical insulation. Another important characteristic of wood is its tensile strength, which is its ability to bend under pressure without breaking. Wood is exceptionally light in proportion to its tensile strength, making it the preferred construction choice for surfaces that take a constant beating such as basketball courts and bowling lanes. Tensile strength is also one of the main reasons for choosing timber as a building material; its remarkably strong qualities make it the perfect choice for heavy-duty building materials such as structural beams.

Of the many construction materials that a person can choose from, wood stands out as a unique and amazingly versatile product. Its aesthetic appeal, tensile strength, insulation qualities, and ease of fabrication enable it to remain a favorite choice for use in an extensive array of construction applications.

Wood Frame Construction

Wood frame construction is one of the most widely used methods to build residential, commercial and industrial buildings. Wood frames are not only very economical to build but are also resistant to extreme climatic variations, and offer residents a high degree of comfort. Added to that, wood frames are sustainable and absorb carbon and offset greenhouse gasses.

Wood frames can be used to construct different styles of buildings and there cannot be any restriction on architectural possibilities when wood is the medium. In order for a wood building to perform its expected functions, it is necessary to construct it judiciously and this can be achieved by using sound construction and erection practices. For instance, wood frames are light in weight and hence it does not require cranes and other heavy machinery for the erection process therefore contributing to the economic aspect of construction.

Foundation for Wood Frames

Based on the type of soil and wood frame loads, a suitable foundation is specified for construction. Footings should be deep enough to avoid frost action in winter. The ground needs to be sufficiently compacted if the structure is constructed in an area where root trees have been removed or filled materials are placed. If the wood frame is constructed over poor soil, treated wood pile with wood or concrete sills can be used.

By and large, two types of foundations are widely used for wood structures: namely concrete foundation and pressure preservative treated wood foundation which are shown in, respectively. Apart from these, free standing pier, piers with curtain walls, and pier supporting grading beam can also be utilized.

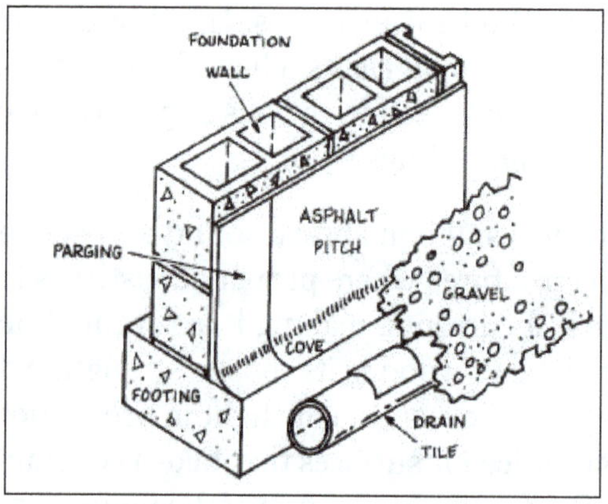

Concrete Foundation.

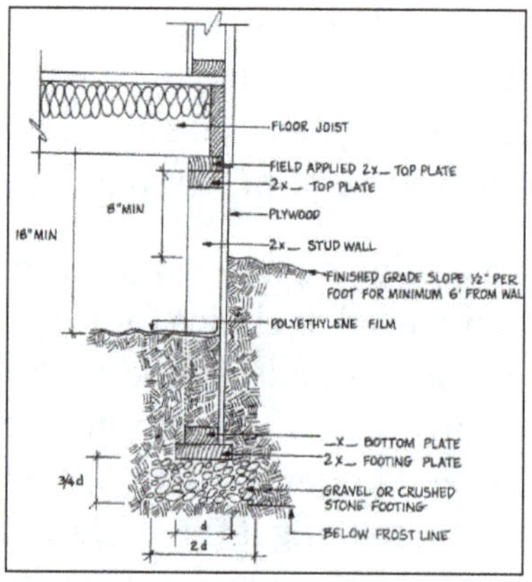

Wood Foundation.

Methods of Wood Frame Construction

After the foundation has been laid, construction of wood frame structure begins. There are various types of techniques that can be used for the construction of wood frames. Suitable construction technique should be selected for a given wood structure.

Platform Frame Construction

It is an easy method and mostly suitable for the construction of houses. In this technique, first-floor joists are covered with sub-flooring to create a surface upon which exterior walls and interior partitions are erected. In the platform system, wall framing can be assembled on the floor and the entire unit can be tilted into its place.

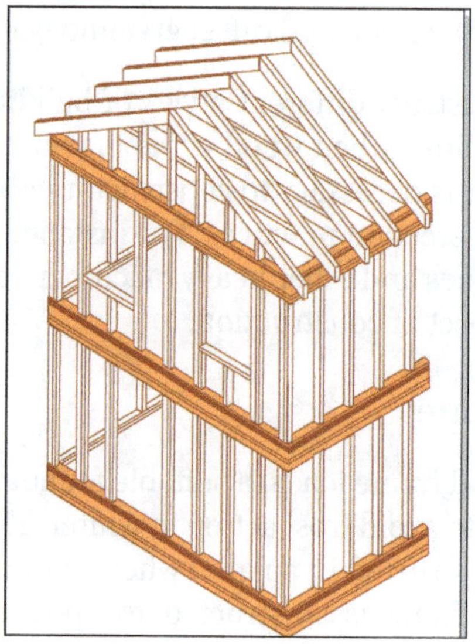

Platform Frame Construction.

Balloon Frame Construction

It is another method of wood frame construction which although is a bit less popular compared to platform frame construction, but is utilized when the conditions ask for it. In this method, exterior wall studs and first-floor joists are supported by anchored sill. The exterior wall studs are continuous to the second floor. The second-floor joists are supported by ribbon strip which are inserted into the inside edges of exterior wall studs.

Balloon Frame Construction.

Plank and Beam Frame Construction

In this framing method, beams, whose ends are supported by posts, are spaced at a maximum of 2.4m and planks are used to cover floors and roofs. The posts provide wall framing and planks act as subflooring and roof sheathing. Wall sheathing is attached to supplementary framing between posts. Roof and floor loads imposed on beams, then get transferred to the posts and are finally received by foundations.

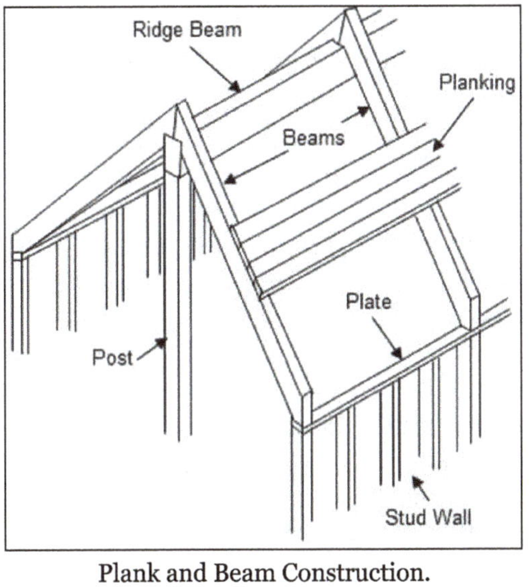

Plank and Beam Construction.

Truss-Framed Construction

In this frame construction method, roof truss, floor truss and metal anchors are used to build strong wood frames. Outstanding advantages of trussed frames are substantial rigidity and wider spacing of roof and floor supporting members.

Truss-framed Construction

Types of Fastenings for Wood Frame

There are various fastening means which can be selected based on the size of wood elements and exerted loads used to attach different elements of wood frames together. Most common fastening techniques are discussed below.

Nails or Combination of Nails, Anchor and Additives

Nails or combination of nails, metal framing anchor, and construction additives are used to fasten framing lumber and sheathing panels.

Nailed Joints

Nailed joints are used when loads act at right angles to nails, but it should be avoided if loads act parallel to the nail.

Wood Protection Measures

Provision of proper means to protect wood is an indicator of good construction practice. Wood frame elements, specifically all the foundation members, shall be protected against decay and termites.

Basic protection measures include the removal of tree root in areas around the structure prior to backfilling, carefully tamp loose backfill to decrease future settlement, and provide certain degree of slopes at foundation and over the building to force water away from the structure.

Finally, all wood structural elements shall be treated based on the exposure to weather and closeness to ground.

Problems Associated with Wood-Frame Construction

For over 100 years, light wood frame construction has been widespread in the United States because it is light, quick, renewable, easily customizable, and does not require heavy tools or equipment. However, contractors, architects, and building owners must contend with several significant problems associated with wood-frame construction.

High and Increasing Costs of Framing Lumber

The cost of framing lumber is reaching record highs. The spike in prices is primarily due to the tariffs on Canadian softwood timber that took effect in November 2017. Lumber prices have also been impacted by wildfires that have destroyed some timberland in British Columbia.

The random lengths framing lumber composite price reached a high of $582 in June 2018; a thirty percent increase over the average composite price in 2017. The National Association of Home Builders (NAHB) reports that framing lumber, including installation, accounts for around 18 percent of a home's average selling price. Consequently, according to Robert Dietz, chief economist for the NAHB, the tariff will increase the cost of a typical newly-built wood framed home by about $9000.

Moisture Problems with Wood-Frame Construction

Wood-frame buildings are susceptible to moisture in their wall cavities. Controlling moisture is challenging because effective methods that stop moisture from entering a wall cavity may also prevent the moisture from leaving the wall cavity. High humidity within a building's cavities is dangerous because moisture can cause wood rot and expensive repairs. High humidity can also lead to the growth of mold, which may cause asthmatic and allergic reactions for the occupants of a building.

Termite Problems with Wood-Frame Construction

Wood-frame construction is prone to termite problems. Termites can damage a structure's integrity and cost thousands of dollars in repairs. In fact, the yearly estimated cost of termite damage and control measures in the United States is $5 billion. Applying termite protection during wood-frame construction is challenging and requires specialized equipment and a trained professional; however, it is essential for maintaining the durability of a wood-frame structure.

Disaster Resistance Problems with Wood-Frame Construction

It is challenging and expensive to construct a wood frame building that has the strength, durability, and resilience to resist storms, tornadoes, flooding, hurricanes, and earthquakes. In earthquake susceptible areas, anchoring a building to its foundation is critical to avoiding structural shifts and the threat of water seepage. In hurricane and tornado areas, contractors must follow strict building code standards for a continuous load path to the ground and ensure a minimum level of resistance to wind loads. Also, to prevent damage to the building's envelope, the windows, walls, roofs, and doors must be missile resistant. Constructing a disaster resistant wood-framed building is doable; however, it can cost 25 – 30 percent more than standard construction.

Wood Frame Construction Lacks Thermal Mass

Wood has low thermal mass. Therefore, wood frame buildings are not as naturally energy-efficient as structures constructed with high thermal mass products like adobe, stone, and Bautex concrete blocks. High thermal mass materials draw in and store heat energy in the day and release the energy at night. The process slows the rate of heat transfer and helps stabilize temperature shifts within a building, which makes high thermal mass products an excellent choice for warm climates.

Waste Problems with Wood-Frame Construction

Wood-frame construction often requires a lot of shaping and resizing of the lumber. The process creates waste and financial loss to the client. However, wood waste generated by commercial and residential construction projects offers the potential for reuse and financial gain. Scraps and cut-offs generated during the trimming and framing constitute a clean waste that can make an excellent feedstock for engineered wood products. To lessen disposal costs and even generate income, builders should contact wood waste processors about setting up drop boxes on site for wood waste scraps.

Sound-Proofing Problems with Wood-Framed Buildings

Constructing a wood-framed building with adequate sound insulation is challenging. Solid and heavy concrete construction, like the Bautex Wall System, provide better noise and sound insulation than lightweight timber. There are several methods contractors can use to achieve sound reduction within wood-frame buildings, including doubling up on the plasterboard or replacing it with a heavier board like Fermacell.

Wood-Framed Construction is Susceptible to Fire Damage

Fire prevention and resistance are crucial and challenging tasks to builders of wood-framed structures. Wood-frame buildings are especially vulnerable to fire damage during construction before contractors have placed fire protection over the frame. The challenge to builders of wood-framed buildings is to stop fires and, in the event of a fire, restrict the spread of flames. Limiting the spread of fire is accomplished by cladding the frame in materials that resist heat and flames and treating the wood with fire retardants.

Swelling and Shrinkage of Wood Frame Construction

Wood swells or shrinks when it gains or loses moisture above or below its fiber saturation point of 28 percent. The fiber saturation point for wood is where all the wood fibers are fully saturated. Above the fiber saturation point, water starts to fill the wood cells. Decay begins if the wood is above the fiber saturation point for a length of time. If the wood is below its fiber saturation point, the wood will shrink. In three, four, and five-story buildings, the effects of shrinkage can affect the building envelope.

Wood-Frame Buildings May Compromise Indoor Air Quality

Wood-frame buildings may also contain volatile organic compounds (VOC), chemicals, and adhesives, all of which will compromise indoor air quality of a building or home. Emissions of VOCs are dangerous because they can cause eye, nose, and throat irritations. VOCs also cause nausea, headaches, and harm to the central nervous system, liver, and kidneys.

Limitations on Designs for Wood-Frame Construction

Limitations in architectural styles and elements are a problem with wood frame construction. With wood frame construction, it is difficult to include large and numerous windows, large spans, and cantilevers and in the design of a wood-framed building.

Steel

Steel is known to have given a structure that no other can when it comes to construction. The durability and potency that steel provides is not matched by wood or concrete. It is getting more and more common for steel to be used in construction, and people are preferring steel due to its various advantages. Steel buildings are built using various frames as well, for example clearspan, modular, single slope etc. Due to the fact that steel is very much easier and less time consuming when it comes to construction, along with its numerous other factors that have proved advantageous, steel, these days, has become the most sought after way of constructing buildings.

Steel buildings are known to have higher resistance due to their strength, and steel construction does not cost as much as concrete or wood construction usually does. It is important to note that there are various advantages of using steel over wood in construction, and due to the number of places starting to incline towards using steel based on its numerous benefits, it has become awfully common. Steel construction is also environment friendly, and can resist natural calamities like earthquakes. Having known these beneficial qualities of steel building construction, let us go ahead and explore the most common types of steel used in construction.

Structural Steel

Structural steel has to meet industry standards for dimensional tolerances and composition. In the United States, steel grades are specified and regulated by ASTM International. Likewise, Europe and Canada have their own regulating bodies and standards. While Leeco Steel does offer EN standard steel plate and CSA G40.21 steel plate.

A wide range of structural steel grades is available, with the most popular being ASTM A572 and ASTM A36. These steel grades along with other structural steel grades are mainly used for building frames of bridges and buildings.

They are also employed in the construction of:

- Freight cars
- Construction equipment
- Truck parts
- Machinery
- Crane booms
- Transmission towers
- Truck frames

Producing and Testing Structural Steel

In order to fully comprehend how structural steel varies from non-structural steel, like the steel used on truck beds, ships, or military tanks, it is important to look at the composition of the structural steel.

Steel can either be made from raw materials, or can be manufactured from recycling old steel. During the process of changing recycled steel into new steel, existing steel is melted down and refined to fit certain specifications. It is a much longer process to make steel from raw materials.

Steel is an alloy containing carbon and iron, both of which are abundantly available but seldom found in pure form. In order to make steel from raw materials, iron is obtained from iron ore, which contains plenty of iron oxides. Most of the iron ore in the United States is extracted from taconite, which is abundantly available in Minnesota. During the extraction process, taconite is ground into a sandy composition, and magnets are used for separating the iron ore (in the form of magnetite) from other substances and minerals.

Although iron is often believed to be strong and hard, raw iron ore is so soft that it can be cut with a knife and some muscle. Iron-based alloys get their strength from the addition of carbon.

An iron-carbon alloy is generally produced by combining coke with iron ore and applying heat until the coke ignites. Coke is a carbon-rich form of coal. As a result of this strong heating, coke drops carbons and adheres to oxygens from the iron oxides, leaving a combination of carbon and iron. This process is called reduction.

Following reduction, around 4% carbon is present in the material, which is subjected to more heating and cooling processes to reduce the amount of carbon, rendering the material stronger and harder. As soon as the carbon content falls below 2.1% of the material's weight, the material becomes steel. In order to make structural steel, carbon should be further reduced until its composition is just 0.05%–0.25%.

The final outcome is a structural steel, which is economically priced, 100% recyclable, and has a high strength-to-weight ratio. Different grades of structural steel are available, all of which vary somewhat in composition. These compositions considerably help in establishing the material required for any specified project.

On the other hand, the steel can be additionally processed — through more heating and cooling treatments and/or the addition of alloys, for example, titanium, molybdenum, and chromium — to boost the hardness. These processes affect the overall brittleness, in most cases, rendering the ensuing material unsuitable for structural applications.

Structural Steel Composition

The following is the composition of two well-known grades of structural steel: ASTM 572 and ASTM A36. While other structural steel grades possess analogous compositions, they may also have extra alloys or have undergone additional processing.

Grade	Carbon	Manganese	Phosphorus	Sulfur	Silicone
A36	0.25-0.29%	1.03%	0.04%	0.05%	0.28%
A572	0.18-0.23%	0.5-0.7%	0.035% max	0.04% max	0.150-0.3%
A514	0.12-0.21%	0.85%	Not specified	Not specified	0.28%

A514 also contains 0.2% molybdenum, 0.48% chromium, 0.05% vanadium, 0.02% titanium, and 0.003% boron.

The presence of additional alloys that boost both hardness and brittleness is the main difference in composition between non-structural and structural steel. In certain cases, the additional alloys are still able to create structural-grade steel; however, in other cases, the steel produced is extremely brittle for use in structural capacities.

Yield and Tensile Strength

Beyond chemical composition, tensile and yield strengths help in establishing the steel grade as well as the overall application.

The highest point of stress is yield strength, at which material will change shape permanently. For instance, when someone jumps off a diving board, the board naturally bends to absorb the energy and weight, but after that person leaps from the board, the board will return back to its original shape. The yield strength of the board would be the point where it bends under energy and weight, and stays permanently bent even after the diver launches.

A vital characteristic of structural steel, yield must have some give to absorb weight. For instance, the yield strength in bridges is the maximum weight that can be tolerated by the bridge before experiencing permanent damage.

Tensile strength refers to the point at which the bent material will break. In the above diving board example, this is the energy and weight it would take to break the board.

The following is a chart that demonstrates the tensile and yield points of three common types of structural steel. These points are either calculated in kilopounds per square inch ("ksi") or in pounds per square inch ("psi"), as indicated below. At times, they are also referenced in megapascals.

Grade	Yield Point	Tensile Point
A36	36 ksi	58-80 ksi
A572	42-65 ksi	0.5-0.7%
A514	100 ksi	110-130 ksi

Yield depends on thickness, but the most common grade is 50 ksi.

These two metrics are very important to engineers as they plan the required material for a specific project.

Structural steelwork is generally used to form a the 'skeleton' frame of a building or other built asset, typically consisting of columns and beams which are riveted, bolted or welded together.

There are a number of different forms of structural steelwork. Some of the more common forms are described below.

Standard Hot-Rolled Sections

Hot-rolled steel is commonly used to form steel beams and columns on construction projects. They are created by passing heated steel between large rollers, which deform it into the required shape, such as; H, I, W, S, C, angles, tubes, and so on.

The standard sections available for use in structural steelwork are given in:

- BS 4-1: 2005 – Structural steel sections. Specification for hot-rolled sections.
- BS EN 10210-1: 2006 – Hot finished structural hollow sections of non-alloy and fine grain steels.
- BS EN 10056-1: 1999 – Specification for structural steel equal and unequal angles.

Common terminology of steelwork sections include:

- Flange: The top and bottom horizontal elements of the 'I'. This resists most of the bending moment experienced.
- Web: The vertical element. This resists shear forces.

- Root radius: The radius of the weld point at which the flange meets the web.
- Toe radius: The radius of the outside edge of the flange.

Typical standard steelwork sections include:

Universal Beam (UB)

This is a beam with an 'I' or 'H'-shaped cross-section that comes in a variety of standard sizes. It is a very efficient form for carrying bending and shear loads in the plane of the web.

The standard method for specifying the dimensions of a standard hot-rolled steel section includes using initials to designate the type of section. For example:

'UB 203 x 133 x 25' – A universal beam of nominal dimensions 203 mm deep, 133 mm wide, and weight 25 kg/m.

Universal Column (UC)

This is a column with an 'I' or 'H'-shaped cross-section that comes in a variety of standard sizes. Unlike a universal beam, the UC's width is roughly equal to their depth.

Dimension specification example:

'152 UC 23' – A universal column of nominal dimensions 152 mm wide, 152 mm deep, and weight 23 kg/m.

Joists

Joists are used to support the floors or roof of a building. A steel joist must be designed to withstand forces and stress while minimising weight, space requirements, and material costs. Different types of joist designs are available based on loading and mounting configurations. Open-web steel joist (OWSJ) consists of two parallel members, or chords, with a repeating, triangular web structure located between the chords.

Channels

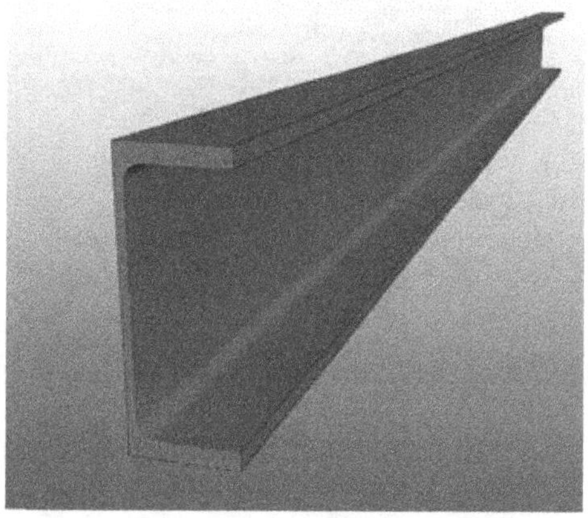

Channels are also known as parallel flange channels (PFC) or a 'C-section'. These are commonly used for columns, lintels above doors, or as a beam supporting floor joists.

Hollow Sections

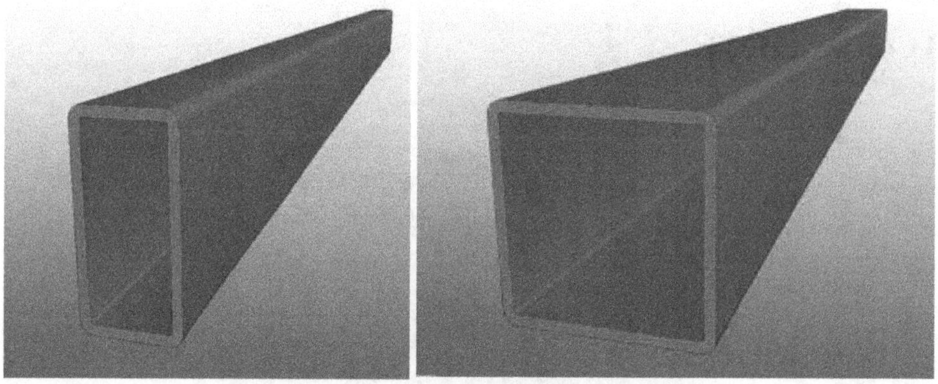

These are typically rectangular hollow sections (RHS) or square hollow sections (SHS). They are used less often than UCs and UBs because they are more difficult to bolt to other beams or members. Dimension specification example:

'250 x 150 x 5 RHS' – An RHS of nominal dimensions 250 mm wide, 150 mm deep, and 5 mm thick.

'250 x 250 x 5 SHS' – An SHS of nominal dimensions 250 mm wide, 250 mm deep, and 5 mm thick.

Flat Sections

Where a connection between two beams is required, flat sections may be used. When welded along the top or bottom length of a beam they can add considerable strength.

Dimension specification example:

'40 x 30 flat' – A flat section of nominal dimensions 40 mm wide and 30 mm thick.

In compound sections, flat sections are often cut smaller as 'plates' to be welded to joists or other steel sections.

Rolled Steel Angle (RSA)

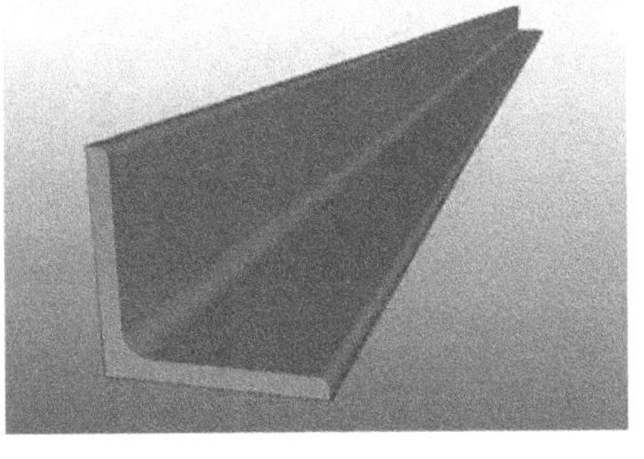

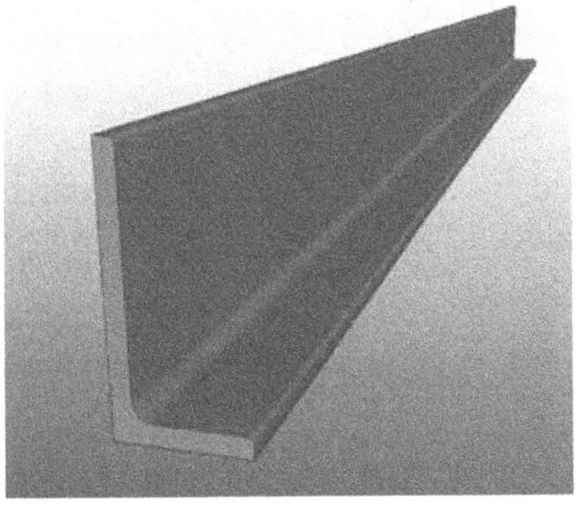

These can be either equal angles or unequal angles and are often called 'L-shape' sections.

Equal angles have legs of the same length and are most often used for balconies, stairs, concrete supports, and platforms.

Dimension specification example:

'50 x 50 x 8 RSA' – An RSA section of nominal dimensions 50 mm wide, 50 mm deep, and 8 mm thick.

Unequal angles have a longer vertical leg. They are used in many of the same applications as equal angles but are often also used as lintels as they have greater loading capabilities.

Dimension specification example:

'50 x 30 x 5 RSA' – An RSA section of nominal dimensions 50 mm deep (longer leg listed first), 30 mm wide, and 5 mm thick.

Circular Hollow Section (CHS)

These are often used as columns and braces steel frame structures. They are listed with their diameter and wall thickness. For example:

'60.3 x 5 CHS' – A CHS section of nominal dimensions 60.3 mm in diameter and 5 mm thick.

Compound Sections

Compound sections are produced by welding together standard sections. They can be used for designs that specify very high loads and long spans that would be unsuitable for standard sections alone. Common variations include:

- Beam or column compound: A standard joist, beam or column section with a steel plate welded above and below.
- Channel compound: These can be either a standard channel welded back-to-back; back-to-back and welded with flange plates above and below; or welded toe-to-toe.
- Gantry girder: These are a combination of standard channel, joist and plate sections welded together.

Open Web Beams

These are commonly used for structures that require long spans with light-to-moderate loading. Deflection is resisted by the relative increase in depth, and structural dead loads can be reduced by the voids that are included in the web.

The different types of open web beam include:

- Perforated beam: A standard beam section with circular voids cut through the web. The voids are spaced to suit the required load.
- Castellated beam: An oxy-acetylene torch is used to cut a standard beam section in two along the web. Each section's projections are then welded together forming a beam that is twice as deep as before.
- Litzka beam: Similar to a castellated beam, but the overall depth is increased further by welding spacer plates to the projects.

Benefits of using Structural Steel for your Building

Without structural steel our world would look quite different. There would be no skyscrapers creating unique patterns against the sky. Buildings would be just a few stories tall and additional square footage would be made up in length and width. Cities would sprawl much farther than they do today. Structures built of materials other than steel would not withstand the extreme weather and seismic events the earth throws at us.

Structural steel makes our world possible, bestowing seven benefits that keep it at the forefront of today's construction industry.

Safety

Safety is the primary goal of any building; steel provides a majority of the safety benefits everyone expects when they enter a structure.

- Steel is non-combustible. It does not ignite nor spread flames.
- It is resistant to rust and corrosion when properly coated.
- It does not mold or mildew.
- It resists splintering and shattering during extreme movement.

When built to code a steel structure will protect occupants and contents from fire, high winds and heavy snow and ice during conditions that would burn, shatter or collapse a building constructed of concrete or wood.

In fact, the safety benefit of steel begins during construction. By using prefabricated building solutions construction time is significantly shorter, meaning less time and fewer reasons for accidents to occur. Reducing or eliminating onsite cutting, forming and welding mitigates the chances for workers to suffer cuts and burns.

Reduced Construction Costs

Prefabricated building solutions provide another benefit of steel - lower costs across the project.

- A reduced timeline results in reduced hours of paid labor.
- A prefabricated steel building solution arrives ready for assembly. Onsite cutting, welding and fastening are not required, saving time and increasing site safety.
- When the frame and envelope are completed quickly, skilled trades can enter and begin work more quickly.
- Stringent fabrication tolerances and a tightly controlled manufacturing environment reduce rework from construction errors.
- A shorter schedule reduces general condition costs across the board.

With fast construction the structure is quickly operational, generating revenues sooner than traditional construction projects.

Future Adaptability

Steel buildings and frames are remarkably adaptable. They are easily expanded either upwards or to any side. Because steel is extremely strong for its weight it can support the added weight of new stories. The overall weight of the structure is still lower than one built with concrete or wood, so the foundation experiences less stress from the added floors.

Besides expanding the building's size, the interior of a steel frame building can be reconfigured with little trouble. Clear span construction provides open space without obstruction created by columns. The area can be configured and reconfigured into a variety of spaces with lightweight interior walls, ceiling systems and movable flooring.

High-Quality Construction

The predictable properties of steel allow designers and fabricators to meet tight tolerances with precision and accuracy. Variation in cutting, punching and rolling that occurs with onsite manual processes is eliminated. Steel members have known strength and dimensions, allowing architects and engineers to predict the feasibility of their designs accurately.

Steel building manufacturers maintain a controlled environment with internal quality control programs dedicated to delivering the expected product. Assembly and erection are performed under controlled conditions at the building site, where the contractor monitors the location and elevation of the structure as well as field bolting and welding.

Serviceability and Resilience

Usability and occupant comfort are critical elements of building operations. A steel building can be designed to eliminate vibration from human, machine or weather movements. Steel exhibits a predictable amount of sway in common conditions with limited movement.

Steel structures are easily repaired, even after extreme damage from high winds, seismic activity or explosion. They resist buckling, distortion and warping under excessive loads.

Design Flexibility

Most of the unique building designs seen today are not possible without steel. Steel is a dynamic material capable of being formed into endless shapes from simple to complex geometries. Its strength allows slender designs not possible in wood or concrete.

Steel building interiors can have floating floors and disappearing walls. Large windows that let in natural light are only possible with a steel frame. Steel frames readily integrate mechanical systems, reducing building volume and energy consumption.

Sustainability

Steel is one of the most sustainable materials in the world. It was green before green was a thing.

- Structural steel made in the U.S. contains 93 percent recycled material, on average.
- 98 percent of all structural steel is recycled into new products.
- Steel loses none of its strength or other physical properties even after repeated recycling.
- The steel-making process has a 95 percent water recycling rate with no external discharge.
- The net consumption of water per ton of steel produced in a mere 70 gallons.

The steel industry reduced its greenhouse emissions per ton by 45 percent since 1975. Building manufacturers and contractors generate minimal waste during fabrication and erection. All scrap is recyclable and can be resold.

Rebar Steel

Functions of Rebar in Construction World Rebar is the short for reinforcing bar. It is also known as reinforcing steel or reinforcement steel. It is tension device made of steel bar or mesh of steel wires used in reinforced concrete to reinforce structure to strengthen and hold the concrete in tension. Its surface is often pattered to form a better bond with the concrete. This is a brief definition on what a rebar is. But when it comes to use there are few basic facts which none should know before buying or implementing.

The basic functions of rebar are:

- *Rebar* has ribs that bind it mechanically to the concrete for better holding of the structure.

- Common rebar is made of unfinished tempered steel, making it susceptible to rusting.
- In high corrosive environment fiber reinforced polymer rebar is used.
- It is usually combined with other steel products, melted down and reformed to gain extra strength.
- Rebar absorbs the shock of an earthquake for certain time before collapsing the building and gives the occupants a "sway" warning and it can help them evacuate the place immediately.
- They are quenched with water at a high pressure so that the outside surface is hardened and the inner cores remain soft which gives the rebar an amazing quality to sway during earthquakes instead of breaking down.
- They are ribbed so that the concrete can have a better grip.
- Coastal regions uses galvanized rebar to prolong the life of rebar.
- They are also used in areas where the environment is corrosive in order to provide no corrosion durability to the construction.
- They when combined with concrete become one of the strongest objects ever and that strength is used in the construction purposes.

Above are all the basic functions about rebar. Knowing all these may help in a construction of a strong stable building which will not fall of easily causing fatal damages. Even if the damages are not fatal many casualties may caused due to collapse of a building. Even if some mega structure is built and it collapses then it will be a huge loss to the economy of the country and the people. Rebar is the basic building block of a building or structure. Before doing anything with the structure one should always be careful on the basic points of a rebar and should get the best rebar for a strong and stable building which will not simply collapse in any worst situation.

Rebar: Reinforcing Steel Bar Specifications

Reinforcing bars are hot-rolled using different steel materials. Most rebars are rolled from new steel billets, but others are rolled from steel debris or railroad rails. Rebars are required to contain some sort of identification that can be used to identify the mill that produced the reinforcing steel bar.

The American Society for Testing and Materials (ASTM) has created a standard identification ruling that all rebars must comply with:

- The number must identify the bar size.
- The type of steel symbol must be noted. For example, "N" means the bar was rolled from a new billet, "W" stands for weldable steel, and "A" designates rolled axle steel.
- The rebar grade identification must be cited: This is either 60 or 75, or metric 420 or 520. The grade indicates the rebar yield strength.
- A symbol identifying the manufacturer that rolled the bar must be included: This is usually a single letter or a plain symbol.

Lower-strength reinforcing steel bars have only three marks that identify the mill that produced the bar, the rebar size, and the type of steel used. High-strength reinforcing steel uses a continuous line system to show steel grade. If the rebar contains two lines, it indicates that the rebar was rolled into the 75,000-psi bars. When a single line is present, it represents a 60,000-psi bar.

Types of Rebar

- Carbon Steel Rebars: This is the most common type of rebar and is sometimes referred to as a "black bar." It's extremely versatile but it corrodes more easily than other types, making it inappropriate in areas that are subject to high humidity or in structures that are frequently exposed to water. Many consider carbon steel rebars to be the best option in all other types of construction, however.
- Welded Wire Fabric: Welded wire fabric (WWF) is made from a series of steel wires arranged at right angles and electrically welded at all steel wire crossings. It can be used in slab-on-ground slabs where the ground has been well compacted. A heavier fabrication of welded wire fabric can be used in walls and structural floor slabs. This is commonly used in road pavement, box culverts, drainage structures, and in small concrete canals.
- Sheet-Metal Reinforcing Bars: Sheet-metal reinforcement is commonly used in floor slabs, stairs, and roof construction. Sheet-metal reinforcing is composed of annealed sheet steel pieces bent into corrugations of about one-sixteenth of an inch deep with holes punched at regular spacing.
- Epoxy-Coated Rebars: Epoxy-coated rebars are expensive and used in areas that will be in contact with salt water or where a corrosion problem is imminent. The only problem is that the coating can be very delicate, so bars should be ordered from a reputable supplier.
- European Rebars: These rebars are typically made of manganese so they tend to bend more easily. They're not suitable for use in areas that are prone to extreme weather conditions or geological effects, such as earthquakes, hurricanes, or tornadoes. They can be cost-effective, however.
- Stainless Steel Rebars: Stainless steel can be used as an alternative reinforcing steel bar with carbon steel reinforcement. Using stainless steel reinforcing bars will not create galvanic corrosion, and it can be a cost-effective solution in areas subject to corrosion problems or where repair is difficult and expensive. These rebars will cost at least eight times more than epoxy-coated rebars, however.
- Galvanized Rebars: Galvanized rebars are 40 times more resistant to corrosion than carbon steer rebars, making them ideal for structures that will be heavily exposed to wet and humid conditions. They're pricey, however.
- Expanded Metal or Wire Mesh Rebars: Expanded metal or wire mesh reinforcement is another good product for concrete. Expanded metal is made by shearing a sheet of steel into parallel lines that are then expanded to form a diamond shape or a square shape between each cut. Expanded metal is commonly used as reinforcement in areas where a considerable thickness of plaster is required, or to reinforce light concrete construction. Wire mesh

reinforcement can be used on sidewalks, small concrete pads, or walkable surfaces that don't receive high live or load charges.

- Glass-Fiber-Reinforced-Polymer (GFRP) Rebars: Similar to carbon fiber, GFRP rebars will not corrode — ever, under any conditions. You'll pay dearly for that, however. These rebars can run 10 times the cost of epoxy-coated rebars.

Rebar Prices and Cost

The cost of rebar can be estimated on a per-foot basis or per ton. The cost of steel can vary from month to month and even from day to day, so if you find a good price, be sure to lock it in ahead of time.

Mild Steel

Mild Steel most common type of steel used in building construction, which is also known as mild steel. It is incalculably strong and durable, and ensures a sturdy built. Due to the strength that carbon steel provides, it is hugely useful in buildings and has proved to be of great advantage. It does not crack when bent, it is immensely flexible, and it is ductile and has great plasticity, along with the fact that it can endure calamities like earthquakes without it causing cracks in the steel. This is the most advantageous factor of carbon steel.

Most Important Mild Steel Properties

- A small amount of carbon makes mild steel to change it properties. Different amount of carbon produces different types of steels. There are small spaces between the iron lattice. Carbon atoms get attached to this spaces and makes it stronger and harder. The harder the steel the lesser the ductility.
- The modulus of elasticity calculated for the industry grade mild steel is 210,000 Mpa. It has a average density of about 7860 kg/m^3.
- Mild steel is a great conductor of electricity. So it can be used easily in the welding process.
- Because of its malleability, mild steel can be used for constructing pipelines and other construction materials. Even domestic cookwares are made of mild steel. It is ductile and not brittle but hard.
- Mild steel can be easily magnetized because of its ferromagnetic properties. So electrical devices can be made of mild steel.
- Mild steel is very much suitable as structural steel. Different automobile manufacturers also use mild steel for making the body and parts of the vehicle.
- Mild steel can be easily machined in the lathe, shaper, drillling or milling machine. Its hardness can be increased by the application of carbon.
- Mild steel is very much prone to rust because it has high amount of carbon. When rust free products are needed people prefer stainless steel over mild steel.

Common Applications of Mild Steel

Here are some examples of where it is used in the world:

- Structural steel
- Signs
- Automobiles
- Furniture
- Decorations
- Wire
- Fencing
- Nails

Stainless Steel in Construction

As stainless steel possesses similar desirable properties to normal steel and eliminates their disadvantages, stainless steel has been widely used as a construction material. The rate of growth of stainless steel use in the civil engineering sector is rapid. It is estimated that in 2006, approximately 14% of the world's total stainless steel production went into the construction industry.

The use of iron in human history can be traced to around 4,000 years BC when early civilizations in Asia and Africa used iron ore to make tools for agricultural purposes. In about 1400 BC, people learned to strengthen iron by heating iron ore and charcoal in simple furnaces.

In early 300 BC, crude steel, which is stronger than iron was made by reheating iron with other metals, however, it was not until the 8th century that cast and wrought iron were first introduce to civil engineering. The earliest application on record is a suspension iron chain bridge in China.

15th to the 19th Century

Britain started to manufacture large amounts of iron in the early 15th century (Swank, 1892). During the 15th century, improvements of furnaces accelerated the development of the iron industry in England; between the 16th and 18th centuries, the lack of charcoal in England gave rise to a large amount of iron import from Sweden, America and Russia.

Europeans used iron for roofing members in the 18th century.

The construction of the clear span 'Iron Bridge' at Coalbrookdale in 1777-79 marked an advance in the use of cast iron.

The industrial revolution in the 19th century led to rapid development of the steel industry as the demand for machinery and transportation increased. The use of modern day steel in Britain began in 1856. The Forth Railway Bridge built in 1890 was the first large scale application of modern day steel in Britain. In the construction of the bridge, riveted and bolted work was used to put together the tapered cantilevered trusses into multiple spans.

20th Century to Present

The mass manufacture and use of steel in the 20th century brought great advance to urbanisation; from street lights to telephones, from typewriter to railways, the world benefited greatly from steel.

The most remarkable application of steel was the construction of the skyscraper which led the architectural world to a new era. Steel frames became commonly used in Britain from 1909 as their strength was proven. Though there were many attempts to develop iron alloys from the mid-19th century, applicable stainless steel was not developed until the first decade of 20th century in the United Kingdom and Germany.

Manganese was among the earliest metals added to steel to increase its heat resistant properties in the 1930s and the new alloy was used commercially in the early 1950s. In the mid-20th century, it was found that steels could be hardened by adding metals such as titanium. Duplex stainless steel emerged in the early 1980s. Today, stainless steel is used extensively in roofing, cables of suspension bridges and flood barriers.

Properties

Chemical Properties

Sedriks defines stainless steel as iron alloys containing a minimum of approximately 11% chromium; this amount of chromium prevents corrosion in unpolluted atmosphere and this is the reason that the steel is named 'stainless'. As its name suggests, the corrosion resistance is one of the most important factors.

However, unlike its physical and mechanical properties which have measured values or derived equations, there are no established chemical equations or simple tests for corrosion resistance. The knowledge about corrosion resistance is therefore solely from recorded experience.

Physical Properties

Only a few physical properties are important to stainless steels used for engineering purpose; specific gravity, electrical conductivity, magnetic susceptibility and thermal conductivity. Stainless steels are more resistant to high temperature than carbon steels are. Though the physical properties are not the most important, they may have an effect on the selection of the material.

Mechanical Properties

Fracture Toughness

the toughness of a material is the resistance to fracture when small defects raise the stress locally to the breaking stress. Biggs points out that the stress to cause fracture is inversely proportional to the length of the crack; the speed of propagation of a crack could reach the speed of sound in the material and this implies that the cracking cannot be stopped once it starts even if the load is removed.

Fatigue

When a metal fractures under repeated stress at a level much lower than the ultimate tensile stress, it is called failure by fatigue. Beddoes & Parr suggest that as fatigue always starts at the surface, great care should be taken during surface treatment.

Fatigue is sometimes overlooked by designers even though there have been incidents of failure by fatigue in recent history. The collapse of the offshore platform Alexander Kielland in 1980 which caused the death of 123 people was due to fatigue failure of one leg of the platform. Fatigue failure may also be one of the reasons for the collapse of the overhead walkway across the atrium of Kansas City, Hyatt Regency Hotel in 1981.

Types

There are over 200 kinds of stainless steel composition and there are new ones being developed each year. Each stainless steel can be distinguished by its response under heat treatment and the mode of fabrication.

Stainless steel can be classified into three main categories based on their metallurgical structure: austenitic, ferritic and martensitic.

The Schaeffler diagram relates metallurgical structure with the composition of stainless steels. The nickel (Ni) and chromium (Cr) equivalents based on weight percentage are the y and x axis of the Schaeffler diagram respectively. The equations of the two equivalents are written as:

$$\%Ni\ equivalent = \%Ni + \%Co + 30\,(\%C) + 25\,(\%N) + 0.5\,(\%Mn) + 0.3\,(\%Cu)$$

where Co is cobalt, C is carbon, N is nitrogen, Mn is manganese and Cu iscopper;

$$\%Cr\ equivalent = \%Cr + 2\,(\%Si) + 1.5\,(\%Mo) + 5\,(\%V) + 5.5\,(\%Al) + 1.75\,(\%Cb) + 1.5\,(\%Ti) + 0.75\,(\%W)$$

where Si is silica, Mo is molybdenum, V is vanadium, Al is aluminum, Cb is columbium, Ti is titanium and W is tungsten.

Austenitic Stainless Steel

Austenitic stainless steels are the most widely used category for building applications. Austenitic stainless steel is considered more corrosion resistant than ferritic and martensitic stainless steels. Austenite is a form of iron that contains trace amounts of carbon at 900-1,400 °C. To retain austenite structure in room temperature, some common austenitisers such as nickel, manganese and nitrogen are added.

Austenitic stainless steel with Molybdenum is resistant to sea water and chloride-bearing solutions; this kind of stainless steel is therefore used extensively in aggressive marine and industrial environments. Though austenitic stainless steels cannot undergo heat treatment, they have tensile strengths of the order of 1,000 N/mm2 and can be used as reinforcing bars with concrete.

Ferritic Stainless Steel

Ferritic stainless steel is given its name because the crystal structure of the steel is the same as that of iron at room temperature. This class of stainless steel is magnetic unless it is heated to above 750 °C.

The Schaeffler diagram suggests that ferritic stainless steel should contain more than 12% chromium and very little Nickel. Though sometimes small amount of other elements such as aluminium, titanium and molybdenum are added, ferritics are considered as binary alloys; therefore ferritic stainless steel has a reduced corrosion resistance and is cheaper than austenitic stainless steel.

The application of ferritic stainless steel in buildings is limited to the interior where corrosion resistance is not so much of a factor. Ferritics do not respond to heat treatment and are more difficult to weld and shape than austenitic stainless steels.

Martensitic Stainless Steel

Martensitic stainless steel has approximately 11.5 - 18% chromium content and has an austenitic structure when it undergoes heat treatment and can be quenched to martensite with an increase in hardness.

Although martensitic stainless steels are extremely strong and tough, Beddoes and Parr point out that the strengthening of the material brings some disadvantages: they are more brittle and are harder to weld and shape.

Precipitation-hardened martensitic stainless steels are formed by adding alloying elements into solid solution at high temperature without the formation of coherent solution, followed by rapid cooling which retains the elements in the solution. The desired mechanical properties are achieved by reheating the material to different temperatures.

Duplex Stainless Steel

Steel with 28% chromium and 6% nickel and containing both austenite and ferrite is called duplex stainless steel. This type of steel has both beneficial and harmful characteristics of the two phases.

By adding other austenite and ferrite stabilisers, the composition of the two phases can be varied. A lot of effort has been put into developing the properties of this relatively recently emerged steel.

Duplex stainless steels are normally used when corrosion resistance and strength are equally important. Duplex stainless steel has been a suitable alternative to carbon steel, other types of stainless steel and nickel based alloys; it has been used for civil engineering applications such as blast and weather walls.

Fabrication

Cutting, Joining and Weldability

When stainless steels are used as structural steelwork, they need to be cut into the required shapes or joined together. The cutting process involves sawing and burning while connecting involves welding and bolting. Today cutting is monitored and carried out by computer generated 'virtual templates'. Welding is carried out in manufacture shops while bolting usually takes place on the construction site.

Air carbon-arc cutting is the use of high-velocity streams of air to remove melted metal behind the arc; it is often used for cutting irregular shapes. Plasma-arc cutting directs gas through an electric arc and ionises it. As the positively charged ions combine with electrons at the steel surface, the temperature rises significantly and this cuts the steel. Electron and laser beams are usually used for cutting thin steel when precision is required. Laser beams are commonly used to drill fine holes.

Making a bolt joint is generally cheaper than making a welded joint. Bolt joints are like pegs in holes and tension and shear both need to be considered. Welded joints require the fusion of the connected faces by heating and allow the use of the cross section of a member rather than having holes for bolts.

There are two types of welds, fillet welds and butt welds. Fillet welds do not require shaping for the edges of plates. Butt welding is used when the full strength of the plates needs to be developed and the weld metal is put across the whole thickness of the plates. Cracks may occur in the weld itself or the parts being connected due to the contraction of the weld when it is cooled down.

Cold Forming and Heat Treatment

The process of increasing the yield strength of stainless steel by plastic deformation at room temperature is called cold forming.

In ancient times, copper and bronze were hardened by hammering to form edges; the cold forming of stainless steel works by using the same principle but with much greater control. Today cold forming is achieved by rolling stainless steel into plates and sheets or drawing them into rod and wire. While the strength of stainless steel increases, the ductility decreases.

Heat treatment is used to change the mechanical, physical and chemical properties of stainless steel. It usually involves heating or cooling at extreme temperatures to achieve the desired level of hardening or softening. There are many different methods of heat treatment. Annealing is the most common method which is used to soften stainless steels to make them easier for cold working.

For ferric and austenitic steel, the annealing process heats them until recrystallization occurs and the size of the steel is increased. For martensitic stainless steel, the annealing process heats them until they are austentic and then cools them very slowly; after the process the steels become ferric with dispersed carbide.

Surface Finishing and Treatments

Polished Finish

Internal and external decorations such as stairs and window trims are usually polished by applying abrasives by means of wheels, belts and pads. During the polishing process, rough cut edges, welds and markings are cleaned and dressed. Electro-polishing is used when mechanical methods of polishing are unable to perform the job, for example chequer plate flooring.

Electro-polishing is achieved by immersing the stainless steel in buckets with hot acids and applying an anode current to the steel by the electrochemical cell. After the process, the surface of the steel becomes smooth and can be easily cleaned.

Coloured Stainless Steel

Coloured stainless steel is widely used in civil engineering applications such as revolving doors, wall panels and lifts. Coloured surfaces are obtained by producing a coloured film through chemical processes. First, the stainless steel is immersed in chromic and sulphuric acids at a temperature a little below the boiling point of the mixture. The colour of the steel is changed from bronze to blue, followed by gold, red, purple and finally to green.

Second, the stainless steel is rinsed with water and the film of colour is hardened by cathodic treatment. The appearances of the final product depends on the starting material: a metallic sheen is obtained for a bright and polished surface and matt colour is obtained if the surface was not originally polished. Corrosion resistance of the stainless steel is also increased by this colouring process.

Applications

Roofing

Stainless steel is widely used as a roofing material. In the UK, stainless steel was initially used for church roofing as it was cost effective. Standing seam and batten roll are two ways of roof laying. Seam joints can be folded and welded to form watertight joints for flat roofs and as a substrate for roof gardens. Batten roll forms a bold line at joints.

The most famous application of stainless steel as roofing material is the roof finial of the Chrysler Building in New York built in 1928-30 which is still intact today. A more recent example is the roof of the Thames Barrier. The flat panels are curved in two directions to make a dimpled shape. The roof of Doncaster Bus Workshop, UK is made of stainless steel with a mosaic pattern to soften the visual compact.

Window Walls

Stainless steels have been used for window walls since the 1930s. The most famous application is the stainless steel cover strip used for the Daily Express, London. Individual casements can be made out of stainless steel facings or roll formed stainless steel. Stainless steel has higher heat resistance than that of normal structural steel and it is therefore used for fire-resisting spandrels and panels.

Building Trim

Stainless steel is a popular material used in building interior decorations such as railings and staircases. Stainless tubes are mechanically or weld joined to form the hand railings of staircases.

Grid and plank stainless steels are used for flooring in raised floors and walkways. Thin plates of stainless steel can also be attached to old floors and staircases for renovation.

Structural Work

Stainless steel was not initially widely used for structural purposes in buildings due to the high cost. Over the last decade or so however, the advancement of technology has made stainless steel much more accessible with reduced costs; furthermore, corrosion resistance and durability have improved.

Today, stainless steel is considered one of the most important materials for structural work. Stainless steel is used for its aesthetical and functional properties. Reinforcing stainless steel bars together with concrete make structures more durable and save time for inspections and maintenance.

Failure

Corrosion

Though stainless steel is known for its high corrosion resistance, corrosion can still occur. There are two main categories of corrosion: general corrosion and localised corrosion.

General corrosion occurs when the entire metal surface is uniformly covered with a film of red rust or green patina. Stainless steel usually does not have visible general corrosion in a dry atmosphere, but at high temperatures, general corrosion may occur. General corrosion occurs when stainless steel is immersed in acids and alkalis. Failure by general corrosion is not very critical and could be determined by immersion test or corrosion literature.

Localised corrosion includes pitting, crevice, inter-granular corrosion, stress-corrosion cracking (SCC), corrosion fatigue and erosion-corrosion. This type of corrosion occurs on particular parts of the stainless steel and is usually at a small scale not easily detected. Failure by localised corrosion is catastrophic as the corrosion develops very quickly. The resistance to the localised corrosion is different for each type of stainless steel and it depends on the alloy compositions. Pitting resistance equivalent (PRE) is written as:

$$PRE = \%Cr + 3.3\%Mo + 16\%N$$

which is a method of describing corrosion resistance. This equation can only be used as a rough reference in the selection of stainless steel.

Seismic Performance

Stainless steel is well-known for its ductility and thus is considered less likely to fail during earthquakes than other building materials such as concrete , although research on the seismic behaviour of stainless steel frameworks is limited.

Though most types of stainless steel are ductile, cold formed steel does not have significant ductility and can only be used in earthquake resistant structures with appropriate measures to ensure that they meet the relevant criteria. Standard deviation of the strengths of members should small and beams should fail before columns.

Sustainable Construction

Recycling and Re-use

Construction accounts for 10% of the GDP in the UK. The great economic and environmental impacts of construction make sustainable construction an important issue. Recycling and re-use of construction materials could reduce environmental damage significantly. As stainless steel is made from ore and the demand for it has increased significantly in recent years, the need for recycling and re-use has become more and more important.

Recycling of stainless steel is economically viable as it contains valuable alloy elements. Today, the production of stainless steel involves melting used scrap with other steel alloys. In the UK, 84% of steel components from building are recycled and 10% are re-used.

Renovation

Approximately 45% of building constructions in Europe are renovated each year and repair and maintenance cost 400 billion Euros. Therefore saving energy for upgrading existing buildings is critical.

Today many buildings are renewed by over-cladding and over-roofing, where a new cover is installed over the existing one to improve energy efficiency, fix damage and enhance appearance. Stainless steel can be used as over-cladding roofing panels and brackets for connecting the old and new structures.

Steel Frame Construction

Steel frame is typically consisting of vertical column and horizontal beams which are riveted, bolted or welded together in a rectilinear grid. Steel beams are horizontal structural members that resist loads applied laterally to their axis. Columns are vertical structural members that transfer compressive loads. It can be used to form the skeleton of a building.

Structural steel framing is typically designed, fabricated and erected in accordance with applicable standards for instance American Institute of Steel Construction (AISC) and Canadian Standard Association (CSA).

Types of Steel Frame Construction

There are various types of steel frame construction which include:

Conventional Steel Fabrication

Conventional steel fabrication involves cutting steel members to the correct length and welding them to build the final structure.

This construction process may be executed on site entirely which require massive manpower.

Alternatively, for best results it can be done in a workshop partially to provide better working conditions and reduce work-time.

Conventional Steel Fabrication.

Bolted Steel Construction

In this technique, all structural steel members are fabricated and painted off-site, then delivered to the construction site, and finally bolted in place.

The size of the steel structural members is controlled by the size of the truck or trailer used to deliver steel elements. Commonly, the maximum length of 6m m is acceptable for normal truck and 12m for long trailer.

Bolted steel construction is substantially fast because lifting the steel members into place and bolting are all the works that need to be executed on construction site.

It is considered to be the most preferred construction approach because the most of the fabrication can be done in workshops, with the right machinery, lighting, and work conditions.

Bolted steel construction.

Light Gauge Steel Construction

Light gauge steel is a thin sheet (commonly range between 1-3mm) of steel which has been bent into shape to form C-sections or Z-sections.

It is broadly common and used for the construction of residential and small buildings. Benefits that light gauge steel construction provide include design flexibility, high construction speed, strong, lightweight, easy to remodel, Recyclable, good Quality (durable and low in maintenance).

Light gauge steel frame construction.

Applications of Steel Frame Structure

Steel frame structure is considerably suitable option for the construction of various buildings and skyscrapers due to its strength, low weigh, speed of construction, large spans construction capability. steel frame structure can be used in the construction of the following structures:

- High rise buildings,
- Industrial buildings,
- Warehouse buildings,
- Residential buildings,
- Temporary Structures,

High rise buildings.

Residential building.

Warehouse building.

Industrial steel buildings (Chemical plant).

Temporary steel building.

Advantages of Steel Frame Structural Construction

- Incredibly versatile
- Environmentally friendly
- Sustainable
- Affordable
- Durable
- Erect quickly and easily
- High strength
- Relatively low weight
- Ability to span large distances
- Adaptability to any kind of shape
- Ductility; when subjected to great force, it will not suddenly crack like glass, but slowly bend out of shape.

References

- Wood: understandconstruction.com, Retrieved 2 February, 2019
- Wood-frame-construction, building: theconstructor.org, Retrieved 22 April, 2019
- Problems-with-wood-frame-construction: bautexsystems.com, Retrieved 25 January, 2019
- 3-common-types-of-steel-building-constructions: nuwireinvestor.com, Retrieved 5 July, 2019

- Structural-steelwork: designingbuildings.co.uk, Retrieved 13 March, 2019
- 7-benefits-of-using-structural-steel-for-your-building: globalspec.com, Retrieved 3 May, 2019
- Types-of-rebars-844455: thebalancesmb.com, Retrieved 19 June, 2019
- 3-common-types-of-steel-building-constructions: nuwireinvestor.com, Retrieved 9 March, 2019
- What-is-mild-steel: metalsupermarkets.com, Retrieved 20 May, 2019
- Stainless-steel-in-construction: designingbuildings.co.uk, Retrieved 11 February, 2019
- Steel-frame-structure-building-construction-24906, structural-engg: theconstructor.org, Retrieved 29 August, 2019

Permissions

All chapters in this book are published with permission under the Creative Commons Attribution Share Alike License or equivalent. Every chapter published in this book has been scrutinized by our experts. Their significance has been extensively debated. The topics covered herein carry significant information for a comprehensive understanding. They may even be implemented as practical applications or may be referred to as a beginning point for further studies.

We would like to thank the editorial team for lending their expertise to make the book truly unique. They have played a crucial role in the development of this book. Without their invaluable contributions this book wouldn't have been possible. They have made vital efforts to compile up to date information on the varied aspects of this subject to make this book a valuable addition to the collection of many professionals and students.

This book was conceptualized with the vision of imparting up-to-date and integrated information in this field. To ensure the same, a matchless editorial board was set up. Every individual on the board went through rigorous rounds of assessment to prove their worth. After which they invested a large part of their time researching and compiling the most relevant data for our readers.

The editorial board has been involved in producing this book since its inception. They have spent rigorous hours researching and exploring the diverse topics which have resulted in the successful publishing of this book. They have passed on their knowledge of decades through this book. To expedite this challenging task, the publisher supported the team at every step. A small team of assistant editors was also appointed to further simplify the editing procedure and attain best results for the readers.

Apart from the editorial board, the designing team has also invested a significant amount of their time in understanding the subject and creating the most relevant covers. They scrutinized every image to scout for the most suitable representation of the subject and create an appropriate cover for the book.

The publishing team has been an ardent support to the editorial, designing and production team. Their endless efforts to recruit the best for this project, has resulted in the accomplishment of this book. They are a veteran in the field of academics and their pool of knowledge is as vast as their experience in printing. Their expertise and guidance has proved useful at every step. Their uncompromising quality standards have made this book an exceptional effort. Their encouragement from time to time has been an inspiration for everyone.

The publisher and the editorial board hope that this book will prove to be a valuable piece of knowledge for students, practitioners and scholars across the globe.

Index

www.ingramcontent.com/pod-product-compliance
Lightning Source LLC
LaVergne TN
LVHW080907210926
848485LV00004B/610